ADVANCE PRAISE FOR

A Solar Buyer's Guide for the Home and Office

"*A Solar Buyer's Guide for the Home and Office* provides a detailed and clear overview of the many solar options available today to power and heat our homes and businesses. If you are confused about where to begin—or whom to turn to for guidance and advice—this book is a great place to start. I highly recommend it."

—GREG PAHL, author of *The Citizen-Powered Energy Handbook: Community Solutions to a Global Crisis*

"Once again, Stephen and Rebekah Hren have delivered an authoritative sourcebook. . . . [T]heir *Solar Buyer's Guide* is a thorough, readable look at how we might turn to the sun to meet our energy needs. It delivers history, explains policy, and untangles the briar patch of the solar jungle. Thanks to this book, the path to sustainability is significantly less perilous: No machete required."

—LYLE ESTILL, author of *Small Is Possible* and *Biodiesel Power*

"Need a clear guide to approaching solar-energy choices? Stephen and Rebekah Hren's new book will help you sift through the marketing hype and make sound decisions for your clean-energy future."

—IAN WOOFENDEN, Senior Editor, *Home Power* magazine, and author of *Windpower for Dummies*

"Most people appreciate solar energy—much as they do motherhood and apple pie—but many do not know about the practical everyday uses of solar power. While they grasp the simple concepts, they are commonly confused by the intricacies of buying solar electricity, heating, and hot-water systems. Hooray for this no-nonsense guide that makes this quagmire comprehensible!"

—JOHNNY WEISS, co-founder and Executive Director, Solar Energy International

A SOLAR BUYER'S GUIDE FOR THE HOME AND OFFICE

A SOLAR BUYER'S GUIDE FOR THE HOME AND OFFICE

Navigating the Maze of Solar Options, Incentives, and Installers

STEPHEN & REBEKAH HREN

CHELSEA GREEN PUBLISHING
WHITE RIVER JUNCTION, VERMONT

Contents

A SOLAR BUYER'S GUIDE FOR THE HOME AND OFFICE

Navigating the Maze of Solar Options, Incentives, and Installers

STEPHEN & REBEKAH HREN

CHELSEA GREEN PUBLISHING
WHITE RIVER JUNCTION, VERMONT

Project Manager: Patricia Stone
Developmental Editor: Cannon Labrie
Copy Editor: Eric Raetz
Proofreader: Helen Walden
Indexer: Lee Lawton
Designer: Peter Holm, Sterling Hill Productions

Printed in the United States of America
First printing September, 2010
10 9 8 7 6 5 4 3 2 1 10 11 12 13

Our Commitment to Green Publishing

Chelsea Green sees publishing as a tool for cultural change and ecological stewardship. We strive to align our book manufacturing practices with our editorial mission and to reduce the impact of our business enterprise in the environment. We print our books and catalogs on chlorine-free recycled paper, using vegetable-based inks whenever possible. This book may cost slightly more because we use recycled paper, and we hope you'll agree that it's worth it. Chelsea Green is a member of the Green Press Initiative (www.greenpressinitiative.org), a nonprofit coalition of publishers, manufacturers, and authors working to protect the world's endangered forests and conserve natural resources. *A Solar Buyer's Guide for the Home and Office* was printed on Joy White, a 30-percent postconsumer recycled paper supplied by Thomson-Shore.

Library of Congress Cataloging-in-Publication Data

Hren, Stephen, 1974-
A solar buyer's guide for the home and office : navigating the maze of solar options, incentives, and installers / Stephen and Rebekah Hren.
p. cm.
ISBN 978-1-60358-261-2
1. Ecological houses--Equipment and supplies--Purchasing. 2. Solar heating--Equipment and supplies--Purchasing. 3. Photovoltaic power systems--Purchasing. 4. Contractors--Selection and appointment. I. Hren, Rebekah, 1975- II. Title.

TH4860.H845 2010
697'.78--dc22

2010026377

Chelsea Green Publishing Company
P.O. Box 428
White River Junction, VT 05001
(802) 295-6300
www.chelseagreen.com

INTRODUCTION

Why Solar Energy?

"Truth is like the Sun. You can shut it out for a time, but it ain't going away!"
—Elvis Presley

"Unless someone like you cares a whole awful lot, nothing is going to get better, it's not."
—The Lorax

We have come a long way over the last few centuries in creating a world with comfortable homes, amazing mobility, convenient appliances, and entertainment options galore. Our human energy and ingenuity has focused on methods and designs for restructuring the physical world to accomplish these ends; compared to a hundred or even ten years ago, we have succeeded beyond what anyone could have imagined. All manner and variety of food, music, art, and entertainment are available at our fingertips or a short journey away. Computers organize our lives and keep us in touch with old friends, freezers store our food, and heat pumps keep our homes temperate year-round. Every day new devices are invented that further increase our ability to create and consume.

What all our amazing accomplishments have not yet achieved is the widespread implementation of means to sustainably maintain our existence. Inarguably we are currently living unsustainably, utterly dependent on finite supplies of fossil fuels that pollute the earth to such a degree as to make it uninhabitable for our children and the millions of other species that call it home. We have reached a critical juncture where we need to turn the bulk of our attention from acquiring new devices that consume dwindling amounts of fossil energy to installing devices that produce renewable energy to power our lives. The health of our planet and the security of nations depend upon it.

Figure 0.1. Almost all homes can utilize at least some solar power, and many have the potential to produce as much as or even more than they consume, like this 2009 Solar Decathlon home built by the University of Louisiana at Lafayette. PHOTO, JIM TETRO/DOE SOLAR DECATHLON.

How do we move from a civilization based on fossil fuels to one based on renewable energy? Fortunately, an answer to this vexing quandary rises every dawn. Solar power is a well understood, universally available source of energy that can provide nearly all of the energy you need—making your office electronics whirr and hum, heating the water for your morning shower, warming the room you're now in. The technology to harness this amazing power has become increasingly efficient and sophisticated, to the point where installing at least one type of solar power system makes sense for almost every home and office. Solar power has both advantages and limitations, and we'll flesh these out in the next chapter—but suffice to say, this type of renewable energy should have a place in nearly all buildings.

History of Solar Power

The power of the sun has been used for millennia to heat homes and start fires, and for centuries to heat water and create electricity. From the bathhouses of Rome, to the layout of ancient Chinese

Figure 0.2. The capture of solar energy has been designed into buildings for millennia, as shown in this Roman ruin of a *heliocaminus*, or "solar furnace." The Romans often used this design in their villas to concentrate solar heat into their living quarters. PHOTO, ZANNER, WIKIMEDIA COMMONS.

towns, to the first commercial solar water heaters over a hundred years ago in Florida and California, solar power has successfully provided meaningful amounts of energy throughout history for the myriad folks who incorporated it into the design and function of their homes and cities. In ancient Greece, those who were ignorant of solar energy were considered "barbarians and primitives." As the ancient Greek philosopher and playwright Aeschylus said:

"Though they had eyes to see, they saw to no avail; they had ears, but understood not. But like shapes in dreams, throughout their time, without purpose they wrought all things in confusion. They lacked the knowledge of houses . . . turned to face the sun, dwelling beneath the ground like swarming ants in sunless caves."

Since the time these words were written in the fifth century B.C.E., the harvesting of solar energy has been greatly refined and expanded. Not only is it possible to heat our homes and our water, but in 1839 the ability to convert sunlight into electricity, called the *photoelectric effect*, was discovered by French physicist A. E. Becquerel.

In 1883, Charles Fritts created the first photovoltaic (PV) cell that used the photoelectric effect to create electricity in a controlled manner, and in 1904 Albert Einstein penned the theoretical explanation of the photoelectric effect for which he was later awarded the Nobel Prize.

Throughout the twentieth century incremental progress was made in PV fabrication, but the first PV cells were extremely expensive. The initial application for these cells was to fuel the space race—they were used to power satellites starting in the late 1950s and early 1960s. Entering the 1970s, research intensified, cells turned into modules, and companies focusing on commercial applications multiplied; prices came down, and useful photovoltaics starting appearing on planet Earth in the form of telecommunications equipment, lighthouses, navy buoys, and other remote power applications.

From the early 1980s through the 1990s, large companies like BP and Siemens got in the game. Photovoltaics became a bit more common, but PV applications stayed "off-grid" except for scattered test sites and a very few "guerilla solar" installs (unpermitted systems tied into the utility network). It wasn't until the late 1990s and early 2000s that utility-interactive PV systems that could "spin the meter backward" started to catch on and multiply as technology improved and electric utilities standardized interconnection procedures and approved safety measures. The past ten years have been a whirlwind of change for photovoltaics, with exponential growth, thousands of new companies, new technology and system components, falling prices, and policy hurdles cleared away. PV systems from hundreds to literally millions of watts are now being built every day and across the globe.

Modern solar water heating has almost as long a history. Many early pioneer households had figured out that putting a cast-iron kettle filled with water in the sun would produce warm bathing water by the end of the day, but it was Clarence Kemp who in 1891 patented the first commercially available solar water heater, which he named the Climax. The Climax was revolutionary for several reasons. First, it could be plumbed directly into existing household plumbing. Second, it incorporated several tanks (typically four) that were painted black and enclosed in a glazed box. Black tanks had been used before, but they rapidly lost heat after sunset and were prone to freezing problems much of the year. The Climax's enclosed box helped ameliorate these problems to such a degree that they were even installed in many colder parts of the country for use during the warmer half of the year.

The Walker water heater appeared at the turn of the century, and marked the first tie-in of a solar water heater with a conventional gas boiler to ensure hot water at all times. The first decade of the 1900s saw a flurry of patents and experimentation with solar water heaters, but one company took the cake. Day and Night solar water heaters exploded onto the scene in 1910, and incorporated separate solar panels with an insulated storage tank. This resulted in hot water, day and night, but freezing was still an issue. In 1913 Day and Night introduced a solar water heating system with a separate freeze-proof heating liquid (containing alcohol) that captured heat in panels on the roof and transferred it to an insulated storage tank. Solar water heating could now be installed even in climates with regular freezing temperatures, and provide water heating all year round. Day and Night did very well, especially in California, but huge discoveries of oil and natural gas in that state spelled doom for these more expensive systems. Solar water heating survived in other parts of the country, notably Florida, until World War II directed most manufacturing toward the military. The energy crises of the 1970s reawoke interest in all things solar and revitalized solar water heating, with many companies drawing on previous designs and knowledge. The industry managed to limp along through the cheap oil days of the 1980s and 1990s, and has staged a major revival since the turn of the millennium.

Solar home heating existed in North America's earliest indigenous architecture. Many dwellings in the southwest from precolonial times, especially those built into south-facing cliffs, made very effective use of solar heating. In colonial times, salt box–style homes common in New England were oriented toward the south, and blocked the cold northern winds with a roof sloped almost to the

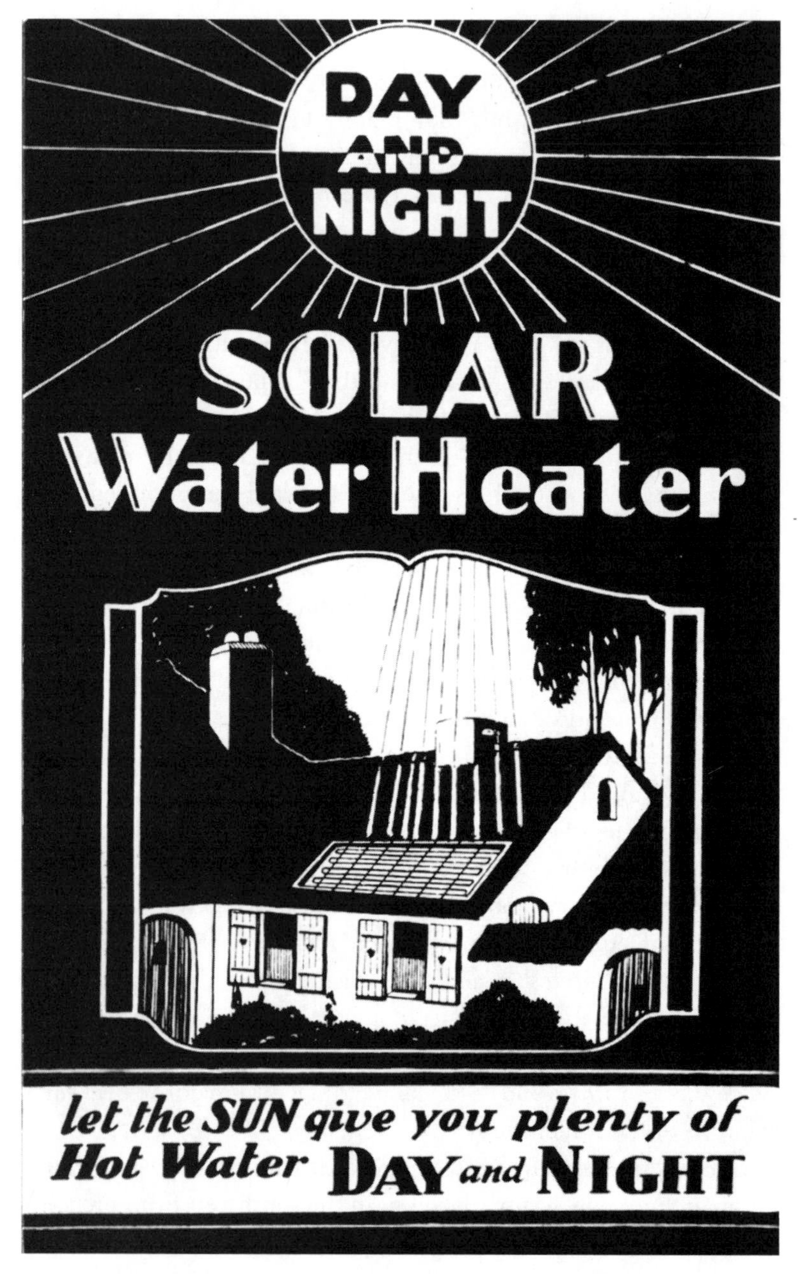

Figure 0.3. Ad for the Day and Night solar water heater circa 1923. From *A Golden Thread: 2,500 Years of Solar Architecture and Technology.* IMAGE © 2010 KEN BUTTI AND JOHN PERLIN.

ground. Modern-day passive solar design begins with the Howard Sloan solar home built in Chicago in 1940 by architect George Fred Keck, who had been experimenting with passive solar design for the previous decade. Sloan did an excellent job of promoting his new home in the media, so that when the country turned its attention to home building after World War II, solar homes were all the rage. A large number of manufactured solar homes were available all over the country. Improvements were made in understanding proper overhang, double-pane windows were introduced, and masonry for thermal mass was incorporated. But the high energy prices of the war dropped off, and interest in solar homes dissipated. The revival during the 1970s energy crises introduced active solar heating, which uses solar panels to warm air or fluid and then transfer heat into the home by bringing this heated air or fluid inside. This made retrofitting solar heating onto existing homes easier, but many of these systems fell out of use when energy prices fell in the 1980s. Improvements and refinements since the turn of the millennium have created a renaissance in solar home heating, and its popularity is growing.

Now at the end of its second century, modern solar power, whether electricity or heat, is available for any building with sun shining down on it. The good news is that solar electricity, like its cousins solar heating and solar hot water, *works right now*—with no need to wait any longer for technological breakthroughs. With sustained technology improvements (and governments everywhere realizing the benefits of domestic energy production) once-prohibitive costs are now within the reach of normal citizens—especially with the help of loans, grants, credits, and rebates.

Solar Power Today: Solar Goes Mainstream

We've burned a lot of fossil fuels over the last few centuries, and undoubtedly we're going to burn a lot more. But an equivalent amount of energy to what was stored in all the fossil fuels the Earth ever contained falls on the surface of the Earth in the form of good old-fashioned sunshine about every six months. And we're finally starting to capture it. Installed solar capacity, defined as the cumulative energy-harvesting capability of all types of in-place solar technology, has been increasing at rates of up to 110 percent a year in this new millennium. But it remains a small percentage of the overall energy mix. As of 2010, only about *1⁄10 of 1 percent* of the energy consumed in the United States is provided by solar devices. Of the 7 percent of U.S. energy coming from what are classified as renewable resources, solar devices provide only a tiny 1 percent of that not-so-big 7 percent. And solar is the most widespread and readily accessible of our renewable resources!

Those miniscule numbers are on the verge of an explosion. All types of solar equipment and technology, from what you see on roofs—like solar electric modules and solar hot water and space-heating collectors—to the auxiliary equipment like racking, inverters, and water heater tanks that keep these systems running, have undergone dramatic and breathtaking improvement over the last few decades. Rural, urban, or suburban, from the southwestern desert to the chilly northeast, one or all of the solar technologies described in this book will work effectively in your home or office. If you take the time to read the following chapters and educate yourself on the options, then you'll see that the world of

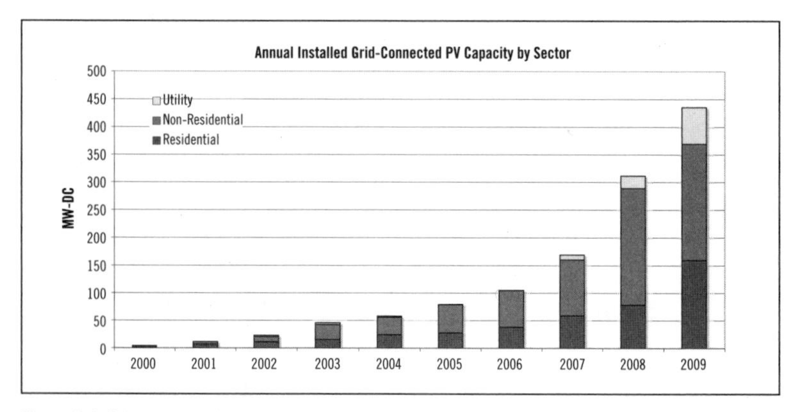

Figure 0.4. Solar power is bursting into the mainstream and growing exponentially, as this chart showing the growth of solar electricity shows. Almost all of new solar power systems are tied into the grid of electric lines, unlike previous off-grid systems that remained separate from utility lines. GRAPH COURTESY LARRY SHERWOOD/IREC.

solar energy is as available, cost-effective, and easy to access as the fossilized technologies that you're used to. And unlike fossil fuel–powered systems, energy from the sun doesn't experience wild price swings, spew out planet-killing and health-destroying pollution and exhaust, come from faraway places whose inhabitants might want to destroy us, or perpetuate our indebtedness. Experiencing the self-empowerment of making your own clean renewable energy is an unbeatable feeling!

About This Book

The book you hold in your hands is a powerful tool, as it will help you gain access to the most versatile and widely available source of renewable energy. Solar power is unique. Unlike other renewable energies like wind or hydropower, sunlight is available in meaningful quantities everywhere on earth. And unlike wind or hydropower, sunlight has the capacity to effectively and efficiently heat things up. Since much of the power our buildings require goes to heating things up like hot water or the buildings' interior, this makes solar power much more useful and accessible as a renewable and sustainable power source. Solar also works well on the small scale of our homes and offices, whereas many other renewable technologies work best on large scales that make them inaccessible or cost-prohibitive to the average person. Much has been written about solar and renewable energies, but we have not come across any modern guides that look at all the solar options comprehensively and untangle the mysteries behind dealing with installers and all the available financial incentives.

We became increasingly aware of the need for a guide such as this after having been involved in all types of solar energy for the last decade. We've designed and built a passive and active solar home from the ground up, as well as retrofitting an existing home to run almost entirely on solar energy. Rebekah's career has immersed her in all things solar. She designs and installs solar hot water, solar power systems, and solar pool heaters. The systems she's installed have varied greatly in design, giving her a very strong background in all types of solar technology. She is a licensed electrical contractor in North Carolina, and a certified PV installer with the North American Board of Certified Energy Practitioners (NABCEP). She teaches solar electricity design and installation classes across the country for nonprofits and universities, and is a certified affiliated PV instructor with the Institute for Sustainable Power Quality. Stephen has taught sustainable building classes that focus on the use of solar energy, and has also taught classes specifically on solar air heaters. He has retrofitted and rehabbed a variety of homes, and has particular experience on utilizing solar energy for heating.

In the following chapters, you'll learn about all the solar options available to use for powering your life. We'll walk step by step through the preliminary analysis of your site, determining which type of solar equipment is right for you and your budget, finding a reputable installer, figuring out what incentives and tax breaks are available, and explaining how each type of system works to give you a full understanding of what is being installed. As you make your own personal transition from a fossil fuel supply to a bright and shiny one, remember that every little bit helps. We also profile inexpensive solar devices to help capture the low-hanging fruit. Solar power is hitting the mainstream, capable of powering almost every part of our homes and offices, and this guide will inform you on all the decisions you need to make to successfully incorporate solar energy into your life. We believe thoughtful, informed decisions are important—so read on and get ready to let the sun shine in!

Resources

U.S. Department of Energy. "The History of Solar": http://www.eere.energy.gov/solar/pdfs/solar_timeline.pdf

Butti and Perlin (1980). *A Golden Thread: 2,500 Years of Solar Architecture and Technology*. New York, New York: Litton Educational Publishing, Inc.

A SOLAR BUYER'S GUIDE FOR THE HOME AND OFFICE

CHAPTER ONE

Types of Solar

What Solar Can Do

Almost without exception, each of our homes and offices could be utilizing more solar energy. The trick is to figure out what's available and what makes sense for your particular climate, location, and budget. If you use hot water, run things with electricity, heat your home or office, use lighting, or cook food, there's a solar device just waiting to be invited into (or more likely on to) your home to displace some or all of the fossil fuels currently doing the job. Becoming thoroughly acquainted with solar resources and available solar technologies is like learning to drive before buying a car. You can certainly shop for solar for your home or business before you learn, but the likelihood of making a good purchase is seriously diminished.

The odds that solar energy will provide *all* of your energy needs are unlikely, although not impossible. It's certainly true that solar energy works better in some places than it does in others, but this doesn't mean it can't work well everywhere (see Figure 1.1). And while less sun can make the financial calculation to go solar less rosy on the face of it, it's equally possible that local incentives can offset all or more of this limited availability. Don't assume that if you live in a cloudier locale that the payback on your solar equipment will take longer—and definitely don't assume that your solar equipment won't work. It will work, and it may pay a great return on your investment. The only way to find out is to figure out your location's solar resource and solar incentives. Before we get into detail on those things, let's take a look at what solar energy can do—and what its advantages and challenges are.

Broadly speaking, solar energy can do three very different things. The most obvious way to use solar energy is *to provide light.* Oftentimes this just means opening up a curtain in the morning.

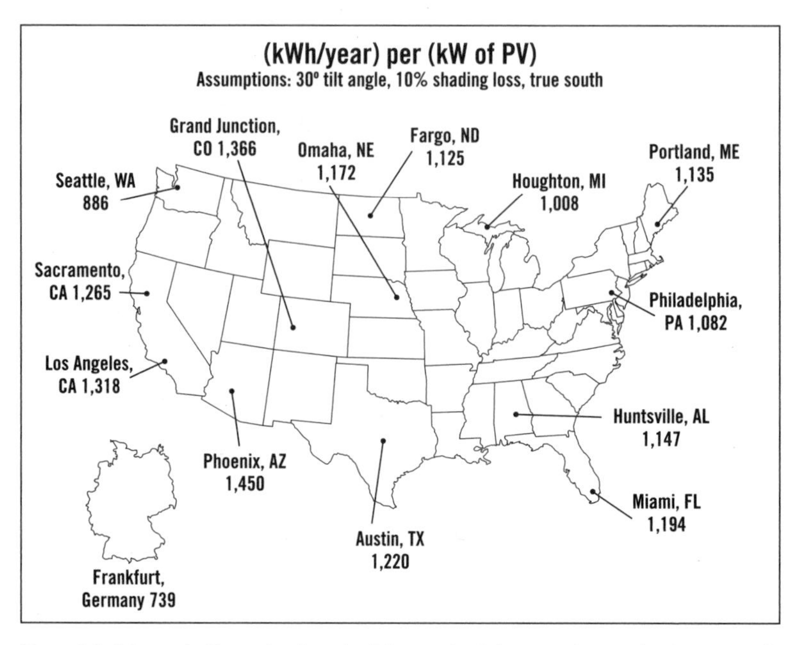

Figure 1.1. Solar production varies throughout the country, but everyone's annual solar resource is nonetheless substantial. This table shows annual average production from a 1,000-watt solar electric system in places throughout the country. Note that everywhere has a higher available solar resource than found in Germany, where approximately ten times more solar has been installed. COURTESY OF SOLAR ENERGY INTERNATIONAL.

But it could also mean installing a daylighting device, something we talk more about in Chapter 8.

A second way to use solar energy is *to heat things up*. We all do this to some degree already. When you sit in the sunshine coming through a window in wintertime or hang up your laundry to dry in the summer, you use solar power to heat things up and accomplish a desired task. A large portion of what we ask nonrenewable fossil fuels to accomplish is to provide heat—for the air inside our home, for our water, or for our food, to name some of the most common uses. All of these jobs make great candidates for solar energy, in part or in whole.

The third way to use solar energy is *to create electricity*. This is a far more complicated task than providing heat, but the results can be much more useful. With solar electricity, instead of just heating things up you can do complicated things like washing clothes, powering your refrigerator, and running your computer.

Q. Is there enough sun where I live?

A. Solar works everywhere. It's that simple. Every single state in the United States has plenty of sun to make solar devices work great (see Table 1.1 for more details). Notoriously cloudy Washington state gets more sunshine per day on average than Germany, although Germany has about ten times more solar electric systems installed than the whole United States. Production may vary throughout the year depending upon prevailing weather patterns and the length of the days, but what's most important is average annual sun hours—not a cloudy month or two. Since all the solar technologies we talk about in this book incorporate other backup power sources, consistent supply is not very important. Instead, the focus should be on displacing as much fossil fuel energy use as possible over the course of the year. Systems installed further north may not produce as consistently in the winter, but can make up for it during long summer days.

Q. Are solar hot water panels the same as solar electric panels?

A. No, they aren't. Solar hot water panels use the sun's heat directly via absorbing radiation to warm water, and solar electric panels use the energy carried in photons of sunlight—via semiconductors—to generate electricity. Solar hot water panels require water to flow through copper tubes inside, but solar electric panels have no moving parts. And while it's the job of solar hot water panels to provide heat and they work better the warmer it is, solar electric (photovoltaic) panels actually work better when they're cold!

Advantages of Solar Energy

If solar energy can do all these things, you might wonder why we bothered with fossil fuels in the first place. That's a good question. But before we get into what makes solar energy a challenge to use, let's look at some of its highlights.

The most obvious benefit of using solar energy is that it's *free* for the taking. Of course, it's not always here when we need it—and sometimes there's too much of it where we don't want it—but in general using solar energy increases our energy independence while decreasing our need to purchase other forms of fuel. This is true both on a personal and a national level.

Second, solar energy is *renewable*. The Earth won't run out of solar energy; more solar energy falls on the Earth in one hour than our human civilization consumes in a year. This will continue for the next few billion years. Once we give solar energy a chance to work for us in our daily lives, all we need to do is maintain the systems we have in place, rather than continuing to scour the globe for depleting fossil energy supplies.

Third, solar energy is *sustainable*. It does not lead to greenhouse gas emissions that disrupt our global climate, or uncontrollable oil spills in the Gulf of Mexico that kill wildlife and ruin fisheries and livelihoods. It does not create toxic waste that will be around for thousands upon thousands of human generations. It is better for ourselves, for our children, and for every other creature with whom we share this globe floating in outer space.

And finally, *solar energy can save you money!* Wild price swings in fossil fuels can rapidly deplete your savings account when you can least afford it, but once solar energy systems are installed, the fuel itself is free. Instead of worrying whether you have enough income to pay next month's heating, electric, or gas bill, once you've installed solar panels your total energy bill can drop all the way to zero! Even if installing solar turns out to be a financial wash, getting clean energy from the sun at a similar price is obviously better than burning more planet-choking fossil fuels.

It may all sound too good to be true—but it's not. Living in homes powered by solar energy is well within our capacity as a creative, hardworking species. Hell, every other species lives within its solar budget! Of course, our civilization has grown to be a little more complicated than most others. But even though over the last century or so our homes have been designed around fossil energy, we've also designed solar energy systems that can do almost all of the tasks we ask fossil fuels to now perform. We'll break down all the different solar devices for you in a minute, but in order to better understand how they work, let's first hash out solar energy's challenges.

Limitations of Solar Energy

Harvesting the tremendous power available from the sun and using it effectively in our complex lives requires some crafty engineering on our part. This is because of the unique characteristics of solar power. Perhaps the most obvious challenge is *availability*. What if we want hot water or electricity when the sun isn't shining? Solar energy is only available intermittently, and some places receive more annual sunlight than others; we either need another energy source to supplement solar energy, or we need to figure out how to store that energy effectively (or some combination of the two). This doesn't diminish the fact that solar energy is a tremendous resource in all climates, from the foggy Northwest to the steamy Southeast.

The issue of availability leads immediately to the issue of *storage*. We can't store sunlight itself, of course, but we can capture this energy in the form of heat or electricity. Doing so, however, might not be easy or inexpensive. Because of the difficulties of storing solar power, it is often most effective to have a supplemental backup source of energy rather than trying to get 100 percent of our energy from the sun. Still, solar energy is capable of providing *most* of the required energy, even in cloudier climes.

Energy from the sun is also *dispersed*. On a bright clear day, around noon, solar energy is remarkably constant at approximately 1,000 watts/m^2 or 3,412 btu/m^2 (a square meter is just a bit larger than ten square feet). The amount of solar energy falling on one square meter in one hour is about equivalent to two ounces of coal, one cup of gasoline, or one cubic foot of natural gas. While this may not seem like very much, remember that there are often thousands of square feet of south-facing roof and wall space on a home receiving sunlight for up to seven or eight hours on many days. The engineering problem faced by most solar panels is to take that dispersed energy and concentrate it where it's needed.

Another issue with using solar power is *climate variability*. Some regions of the country receive more constant sunlight than other regions. Areas with more cloud cover are more subject to the problems of availability and storage than sunnier places. Also, especially for hot water systems, colder regions may require more complex systems to ensure they function properly during freezing weather. These increases in complexity can mean more expensive systems—but with the many incentives in place, even these more complex hot water systems can be affordable and still capture needed energy from the sun.

More complex systems can make the final issue, *affordability*, harder to achieve. For most solar devices, the cost (thousands of dollars for solar hot water and heating, ten thousand or more for solar electric) is borne upfront when the system is purchased and installed. The energy to run the system is, as previously mentioned, free. Because of the large up-front cost, solar energy systems often appear to be quite expensive—but when the cost is averaged out over the system lifetime (and especially when financial incentives are taken into account) these systems often compare favorably with conventional fossil fuel–based technology.

Different types of solar energy collection devices also differ in their *efficiency*, or their ability to convert sunlight into the energy needed to perform the desired task. These differences in efficiency make payback variable among different types of systems. Fortunately, the hurdle of affordability is being lowered by a variety of measures such as state, federal, and utility rebates or tax credits; installer financing; and green-power initiatives. We'll get into the nitty-gritty of all these financing options in Chapter 3, "What's Appropriate for Your Budget."

Types of Solar Panels

Walking past a neighbor's home, you may notice one or more large black rectangles on the roof or walls and suspect that it's some sort of solar device. From the ground, differentiating between the various types of solar panels can be difficult. A typical solar hot water panel, solar electric panel, or solar air heating panel will be a large black rectangle, often two or more together. We'll get into more detail later, but for now let's take a minute to go over some of the more obvious differences in these three main categories of solar panels.

Solar Electric Modules

When someone mentions solar power, most people think about solar electric—or photovoltaic (PV)—panels (more correctly referred to as *PV modules*). These come in a range of sizes, from the tiny rectangle on a calculator to the common roof-size module of approximately three by five feet (with many variations coming from different manufacturers) which rest in frames about one inch thick. They are usually arranged in an array comprised of anywhere from two to several hundred modules, often on roofs but (because electricity is easy to transport) also out in yards or fields. Solar electric modules take the energy (but not the heat) coming in as photons from the sun and use that energy to excite electrons. The solar energy imparted to the electrons from the sun is harvested as electricity. Excited electrons flow in wires from the solar array

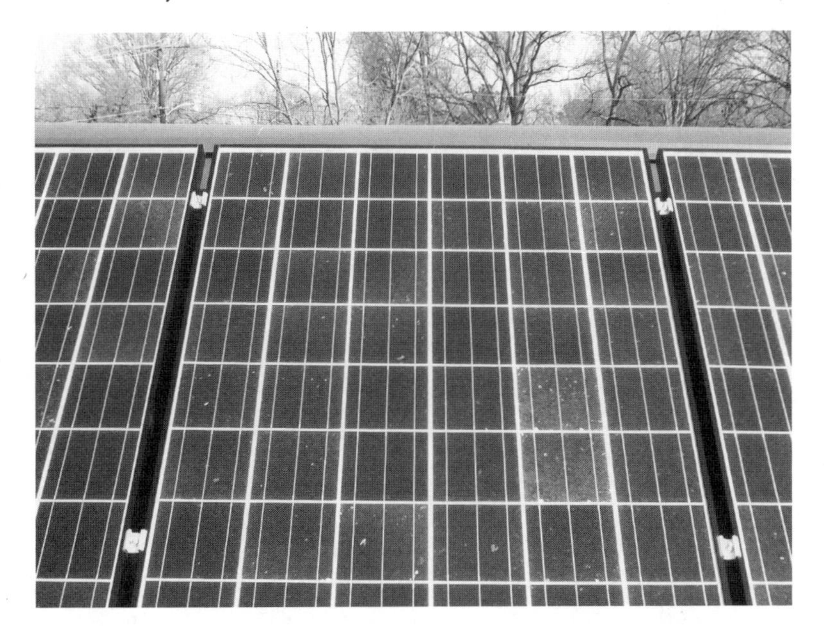

Figure 1.2. A solar electric module, also called a photovoltaic (or PV) module, creates electricity from sunlight. Typical residential PV modules are about three by five feet, and multiple modules are grouped together to form an array.

through a circuit that enters your home, where an electronic device called an inverter matches the voltage and current coming from the solar array to the other circuits in your house. The solar electricity can then be used to power your home, and used to "feed the grid" and spin your utility meter backward.

Solar Hot Water Panels

One of the most effective ways to use solar energy is to heat up water, a process often four times as efficient as generating solar electricity. The payback on solar hot water systems can be quick even in colder, cloudier climates, so sensibly these have historically been more common residential systems than solar electric. Hot water is difficult to transport without losing its heat, so it must be created near where it's used. If you see a pair of large (four by eight or four by twelve) glass-covered black panels four inches or deeper side by side on someone's roof, chances are they're for solar hot water. A system with multiple panels might provide energy for radiant heat as well. Another less common type of solar hot water system uses evacuated tubes instead of panels; if you see a dozen or two elongated glass tubes suspended on a rack on someone's roof, that's also a solar hot water system.

Solar energy is also employed to heat up the water in swimming pools. Heating pool water can use a huge amount of fossil energy, because the heat is constantly being wicked away by the surround-

Figure 1.3. Left, a typical solar water heater panel consists of a glazed insulated black box with copper tubing inside that captures heat from sunlight—the box is generally four by eight or four by twelve feet. Right, this flat-plate solar water heating panel contains small-diameter copper tubing with heat-accumulating fins attached, painted black. The fluid running through the tubing accumulates the sun's heat. RIGHT PHOTO COURTESY SUNEARTH INC.

ing ground or air. Fortunately, solar pool heaters are easy to install and extremely effective, especially when combined with an insulated pool cover. A solar pool heater will generally have at least four long flat plastic panels (four by eight or four by twelve feet) instead of the glass-covered panels used for domestic solar hot water.

There's a great variety of solar hot water systems, and we'll flesh out how these work and the various choices in Chapter 6, "Solar Hot Water."

Solar Air Heater Panels

In the last couple of years active solar air heating panels, often just called *solar air heaters*, have been increasing in popularity. The most common place for a solar air heater is on a south-facing wall, although they can also be installed on a roof (with added difficulty and expense). They are similar in size to solar hot water panels. Solar air heater panels are warmed by the winter sun. Air from an adjacent room is pulled into the bottom of the panel through a vent by means of a small PV panel and fan (or a plug-in fan), heated up, and then blown back into the room through a vent at the top. Generally, solar air heaters are most effective in colder climates,

Figure 1.4. Left, solar heating can be retrofitted onto a building using active solar heating systems, which can employ solar water heater panels like those pictured in Figure 1.2, or solar air heaters like the ones pictured on the lower half of the roof. They are similar in appearance to solar water heaters, but move air instead of a fluid. Right, this picture of a solar air heater shows the baffles that allow for air movement and the embedded PV module that powers the fan. PHOTO AND ILLUSTRATION COURTESY OF YOUR SOLAR HOME, INC.

Figure 1.5. Modern homes designed to let in the sun through south-facing windows in winter, called passive solar homes, have been around for many decades. Overhangs shade the windows in summer. This house also has solar electric and solar hot water panels on its roof. PHOTO COURTESY OF ALICIA RAVETTO.

where the heating load is more substantial. The sunnier it is in the winter, the better, of course.

Solar electric, hot water, and air heating panels make up the three main categories of what are often referred to collectively as solar panels. It's certainly possible to have all three at the same time. In fact, if you have the need and the room, we highly recommend it! Don't be surprised if once you get one kind of solar panel installed, you start hankering after more. Solar power in action is amazing stuff, and collecting various solar systems makes a great hobby. And not many other hobbies save money and the planet at the same time! But these three solar options are not the only way to utilize solar power.

Passive Solar Design and Other Solar Gizmos

Besides solar electric, hot water, and heating panels, there are many other ways to use solar energy effectively, and we want you to know about them all. The major omission so far is using solar energy to passively heat up your home through south-facing windows, often referred to simply as *passive solar*. If you are considering adding or replacing windows on the south side of your home, there are several important considerations to effectively achieve this. We discuss these in Chapter 7, "Solar Heating."

Another category of solar devices are smaller but usually very effective at what they do. There are solar-powered gadgets that can charge your batteries, get the heat out of your attic, let light into your dark rooms, and even bake bread! Many of these will make great additions to your solar empire. We discuss these gizmos in more detail in Chapter 8, "Everything Else Under the Sun."

CHAPTER 2

What's Appropriate for Your Site

Solar Availability for Your Site

Free energy from the sun streams down upon us, and we need to take advantage of it! But to make an investment in solar worthwhile, you must have sufficient access to that free energy. Neighboring buildings, trees, and terrain can all be an impediment to sufficient sun. After all, you can't live a life powered by the sun in the shade! Even if your home and yard are completely shaded, there are still opportunities to invest in solar energy (and especially solar electricity) through things like green-power programs (see the section "What If My Solar Window Stinks?" for more info on green power on page 24).

Q. If I install solar, will it make my house look ugly or make it harder to sell?
A. No and no. Solar installers today work hard to install systems that blend in seamlessly with your roof contour and color, and don't stick out. Plus, the *Appraisal Journal* reports that for every one thousand dollars saved in annual utility expenses, a home's value increases from ten to twenty-five thousand dollars.

Q. Are these systems reliable? Do they need maintenance?
A. Solar systems need very little maintenance, not usually more than a once-a-year checkup. Dry and dusty regions might need panels hosed off a few times a year. Your system will be warrantied, and if installed correctly will last for decades.

Q. I've already got electricity, hot water, and heat—why would I need to buy this?
A. So you know your power comes from clean renewable resources.

To analyze a particular location such as your south-facing roof space or a sunny spot out in the yard, you'll need to take stock of the six primary factors that affect any site's unique solar availability—its solar window. Doing so before bringing in installers to get project bids ensures that you understand what the potential limits and complications are for any solar installation at your site. These six factors are:

1. Existing vegetation and its potential growth
2. Average seasonal insolation and climate extremes
3. Position on Earth (latitude) relative to the sun
4. Orientation of your home and other existing structures relative to solar south
5. Terrain such as mountains and other potential obstructions such as complicated roof designs
6. Available area or square footage

Fossil fuel–based systems pollute, result in energy dependence on rogue states, and leave the Earth a worse place. Solar energy has none of these issues, and you can actually save money on your energy bills in the many places with great financial incentives.

Q. How much is all this going to cost me?

A. Good question! We'll spend a lot of time answering this in detail in the next chapter, because the issues of payback, financial incentives, and performance are detailed and (frankly) a little convoluted. To get it out there, the up-front cost of solar water heating is typically in the three to nine thousand dollar range, a solar electric system is typically ten to forty thousand dollars, and solar heating can range widely from one thousand dollars on up, depending on local climate and the size of the system. Keep in mind that some, and potentially all, of these costs can be recouped through tax credits, incentives, savings on fuel costs, and the like.

HERE ARE THE THREE MOST IMPORTANT FACTORS IN CHOOSING THE SITE FOR A SOLAR DEVICE.

1. There must be sufficient sunshine! Look for little to no shade from 9:00 A.M. to 3:00 P.M. on average, year-round. Solar electric systems perform poorly in any shade, though solar thermal systems (systems that generate heat for hot water or interior heating) can handle a bit of dappled shade from leafless deciduous trees in winter.
2. The azimuth or direction the device faces. Generally south-facing for the northern hemisphere, although any direction other than north can work—especially west for solar thermal systems.
3. The tilt angle of the device or of the roof it will be installed upon. This is not critical for most systems as long as they are facing generally south (within about 30 degrees of "solar" south; see the definition in the "Defining Terms" sidebar). PV arrays can be installed at any angle from flat to perpendicular. If facing east or west, a lower tilt angle is usually better to catch sun during more of the day. Solar thermal systems should not be installed flat if they are being used to provide heating during the winter.

Getting Acquainted with Your Solar Window

If you haven't already, it's time to take stock of where your home, building, or office is in relation to the sun. When you get up in the morning, and over the course of the day, watch where the sunlight comes in and where it goes. Pay attention to where it is at midmorning, midday, and midafternoon. Which exterior walls are shaded? When does the roof get sun in the morning and when does it stop in the evening? What surrounding structures cast shadows on which part of your home, and at what times of the day? Additionally, which parts of the building shade other parts, such as dormer windows, vent pipes, and chimneys?

As the seasons change, many of these things will gradually change with them. When the sun sinks into the sky during winter,

DEFINING TERMS

Insolation or **solar irradiation** is the amount of solar energy that falls on a particular area for a specified period of time. Insolation is commonly given in watt hours per square meter (wh/m^2) or in hours of sun per day. **Peak sun hours** are the number of hours per day that a site is receiving the equivalent of 1,000 watts/m^2. **Solar south** refers to the direction of the sun halfway between sunrise and sunset on any given day.

All of the factors listed above help determine what is referred to as your **solar window**, a description of how much solar access one particular spot has available over the course of the year. Different parts of your home and yard will have different solar windows, and finding the one with the maximum amount of solar access will make sure your solar investment pays off.

Solar thermal is used in this book to refer to solar equipment that is intended to transform solar energy into heat. This can be heat for hot water, heat for the home interior, or both simultaneously.

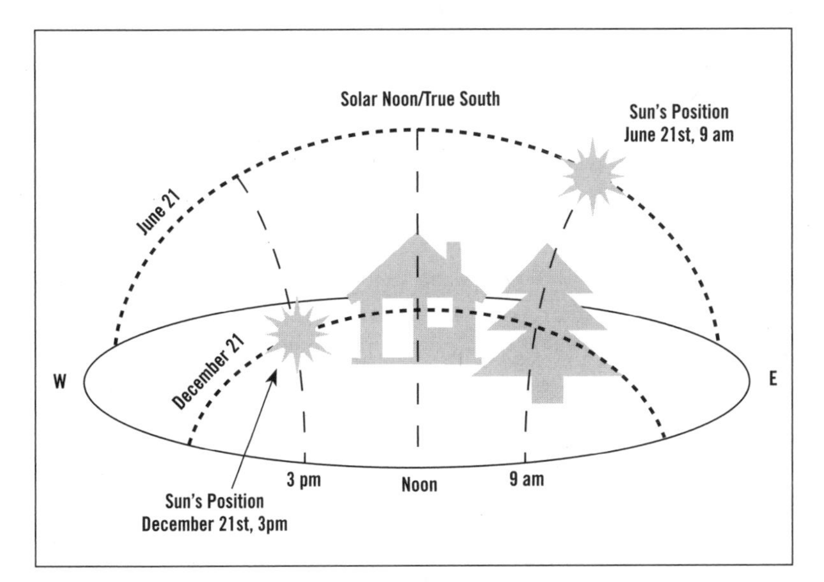

Figure 2.1. Your location's access to the sun, referred to as its solar window, changes throughout the course of the year, making a solar analysis of your site the first step in determining the possibility of using solar energy.

the number of obstructions will likely increase. On the other hand, trees will lose their leaves. Consider how these changes would affect efforts to access sunshine at that time of year. In midsummer, pay attention to the unexpected solar path that starts with the sun to the northeast at dawn, high up overhead at midday, and setting in the northwest at dusk. Each time you watch the path of the sun over the course of the day, remember that it will follow this exact same path on the other side of solstice, so that May 21st will be the same as July 21st, April 21st the same as August 21st, and so on.

Much of what you discover about the position of the sun over the course of the year may be counterintuitive. Since the Earth's axis is at a fairly steep incline (around 23 degrees), and in addition to spinning like a top it's also circumnavigating the sun, your relative position to the sun is constantly changing. Studying the intricate dance between your site's unique solar window and the always-moving sun before trying to harness it will decrease the likelihood of missteps.

The Solar Resource

The amount of power (or irradiance) a spot on the Earth receives from the sun changes over the course of the day and year, because the angle of incoming sunlight through the atmosphere shifts, as does the amount of cloud cover (or what can be called microclimate). Although every individual spot on Earth receives the same total number of hours of *daylight* per year, the energy (also known as insolation or solar irradiation) a specific spot receives varies widely based on the local microclimate. This specific amount of energy can be quantified in peak sun hours, and is measured in blocks of 1,000 watts/m^2. A chunk of energy adding up to 1,000 watts/m^2 for one hour equals one peak sun hour (for example, 200 watts/m^2 for one hour in the morning plus 800 watts/m^2 for one hour in the early afternoon equal one peak sun hour). The National Renewable Energy Laboratory has been collecting data from weather stations for decades, and provides average peak sun hour numbers for different regions.

For solar resources used year-round, such as solar hot water and solar electricity, it's fine to just look at the peak sun hour number averaged over the entire year. For solar heating purposes, you need to separate out individual months, and we go into detail on that in Chapter 7. The next time you hop on the Web, take a minute to visit the NREL Web site (http://rredc.nrel.gov/solar/calculators/PVWATTS/version1/) and find the numbers for the research station nearest you.

Analyzing Your Solar Window

What good will this "average peak sun hour" number do you? First, it allows you to compare your solar resource with other regions of the country, to get an idea of what to expect from any potential solar equipment. It also gives a solid foundation on which to base conversations with solar installers and gauge their knowledge and trustworthiness. Any seasoned installer should be able to converse

Figure 2.2. A solar site analysis for the entire year can be done with a variety of tools, perhaps the niftiest of which is the Solar Pathfinder. A clear dome reflects any potential obstructions and projects them onto the path of the sun for each individual month.

with ease about your region's solar resource, how that information fits into your location's solar window, and the future performance of any solar technologies you have installed.

But the main reason for defining your sun hours is to combine that number with your unique solar window to understand the potential supply of solar energy for your site. With this information, you know how much solar energy is available at your site at various times of year, and you can then estimate how much energy a particular solar device will provide. You will already have an excellent idea of whether solar equipment will work for you, and this will make conversing with potential installers much easier. Many solar installations run into thousands or tens of thousands of dollars, so employing the same diligence as with other large purchases like cars or homes will greatly smooth out any difficulties.

The efficiency of a system depends first upon the available solar energy as described above—which is partially a factor of shading and partially a factor of precise location. Systems lose and gain power depending upon how perpendicular they are to the sun's rays at any given time. So installers will note the azimuth of a site (how closely it is aligned with true south). They will also note the tilt angle of a fixed mounting structure, like a roof. The azimuth and tilt angle will be used in calculations along with peak sun hours to predict how much energy a particular solar device will deliver. See Table 2.1 in the chapter on solar electricity for comparisons of different tilt angles, insolation, and production numbers.

Table 2.1 Comparing PV Array Energy Production Annually at Different Tilt Angles & Azimuths

Size of PV Array	Location	Facing True South (5 Degree Tilt)	Facing True South (30 Degree Tilt)	Facing East (30 Degree Tilt)
5 kW	Durham, NC	5,722 kWh	6,566 kWh	5,308 kWh
5 kW	San Diego, CA	6,845 kWh	7,493 kWh	5,798 kWh
5 kW	Pittsburgh, PA	5,130 kWh	5,553 kWh	4,642 kWh
5 kW	Omaha, NE	5,763 kWh	6,544 kWh	5,192 kWh

Fortunately for us, engineers have designed amazing and effective solar site-analysis tools to make calculating the solar resource quick and easy for a particular location. On the first visit to your site, any solar installer worth their salt should do some preliminary analysis of your solar window using one of these tools. The most common of these nifty little devices is called the Solar Pathfinder. The Solar Pathfinder works by positioning a half-globe over an outline of the path of the sun. By leveling this half-globe and aligning it with true south, the reflection of any potential obstructions at any time of day for any month of the year can be outlined on the sun-path chart (see Figure 2.2). Thus a reading can be made in one instant for a year's worth of changing shadows.

Enterprising engineers have even written smart phone applications—homeowners with iPhones can download an application called the SolmetricIPV for less than twenty dollars that will do a decent solar site analysis. On the opposite end of the spectrum is the Solmetric Suneye, a digital tool that costs upwards of one thousand dollars. Watching a visiting installer use one of these tools (or downloading an app for your smart phone) will go a long way toward bettering your understanding of the sun's movement over the course of each day and each season.

Altering Your Solar Window

Since for the most part we have not planned our homes, buildings, and landscapes with the idea of solar accessibility, it's not surprising that most of our solar windows will be less than wide-open. But it is also important to remember that we inhabit a human-made and human-altered landscape, including not just the shape and orientation of our buildings but also the placement of trees and other potential obstructions. What may seem like a permanent blockage of the sun can sometimes be manipulated and altered—just as the original landscape was. If you're considering a large solar addition to your home, it might make sense to do some preliminary opening up of your solar window so that you get the most out of your investment.

Roof Obstructions

Sometimes roof space will be perfect for solar except for some obstruction such as a chimney, a plumbing vent, or a small architectural flourish (such as a small window into the attic) that either shades the roof or significantly interrupts the roof space. For instance, our house contained an old chimney from the original coal-fired furnace. The existing gas furnace was vented out of this chimney, but before we reroofed we made the decision to take the chimney down and vent the furnace through a much smaller metal pipe. This made the job of reroofing easier, it removed potential shading, and it also removed a large thermal bridge that conducted interior heat to the outside, thereby improving the energy efficiency of our home.

If you're considering large solar additions to your home that may require tens of thousands of dollars of investment, spending a few thousand on clearing the roof beforehand may be money well spent. Not only will it open up the maximum amount of space to solar equipment, but it is likely to make the job of installation easier since racking and panels will not need to be altered or disrupted. Small obstructions like plumbing or attic vents can often be moved quite easily, and even larger ones may not be overly costly to move, depending on the degree of their obstruction and the size of your solar project.

Landscape Obstructions

As a couple trying to make the world a sustainable place to live in, it's painful to have to recommend some judicious pruning or harvesting of landscape obstructions such as trees. And yes—by harvesting we do mean cutting them down, hopefully to be used for one of the many purposes that trees are great for, such as lumber or firewood. Large trees near your house not only disrupt your and your neighbor's solar window—often to the point of making it impossible to utilize solar energy—but they also ruin infrastructure such as sidewalks and roads, gas and water lines, and the foundations of buildings. They can also be a hazard during fire season and cause mayhem during large wind events, something we experienced firsthand during a hurricane. Remember, trees grow! New medium-sized trees (such as the amazing variety of fruit trees) can replace larger ones, and provide summer shade for the yard and walls of a house without ruining your roof's (and your neighbor's) solar access and causing the other dangers that large trees pose. If we are going to create a world powered by the sun and renewable energy, we have to be able to access the sun. If we remain in the shade, we will forever be dependent on fossil fuels—and the resulting climate disruption will likely destroy many of the forests of the world. Bear in mind that it may be necessary to harvest a tree or two to save the forest. And while harvesting large trees can be costly, it will make the payback on any solar investment much quicker, while simultaneously protecting your home and family from danger during storms. And the new fruit tree you plant will bear lots of delicious fruit!

Trees come in two types—deciduous and evergreen. While solar electric systems are very sensitive to even small amounts of shade, solar thermal technology can still function at a reduced level when the sun is only partially blocked, such as through the branches of deciduous trees in wintertime. As the sun sinks toward the horizon during winter, make a note of what trees will block your solar resource—and whether they are deciduous or evergreen. And keep in mind that tree size is not static; trees that are close to any solar installation can shade them out in a few years.

What if the trees blocking your solar window aren't yours? Obviously, if you just go next door and ask your neighbors to cut down their trees, you're unlikely to get a positive response. Feel them out first, and let them know that you're thinking about investing in a solar system. If the opportunity arises, consider offering to pay for harvesting any offending trees and planting some smaller native or fruit trees instead. If this seems impossible, or they refuse you outright, don't despair. It's still possible to access solar electricity through green-power programs; we'll discuss this more on page 24 in "What If My Solar Window Stinks?"

It's also possible that future construction could block out your solar window, possibly after you've spent tens of thousands of dollars. See the "Solar Access Laws" sidebar, and visit the DSIRE

Web site (see "Resources" at the end of this chapter) for more on your state's solar laws.

Aesthetic Concerns

Many folks are fascinated with solar energy and want to figure out how to integrate it into their lives, but aren't particularly excited about altering the look of their house or other building with a few solar panels. It depends on the physical characteristics and orientation of your solar window, but for those with aesthetic concerns about solar panels the options break down as follows:

- Ask the installer to use a part of the roof that is flatter or more hidden for placing the collectors or modules.
- There are color choices for frames, rails, and other structural components of systems, as well as color choices available for solar collectors themselves; make sure your installer knows your preferences.
- Consider placement on a garage or other out-building, or a ground- or pole-mount in the yard.
- If your roof needs replacement, consider building integrated options, such as solar electric laminates that stick to metal roofing—or even solar shingles, which haven't been common but are increasing in market availability.

Solar Access Laws and the Rules of Your Neighborhood

Before contacting a solar installer, you should also check into your **CCRs**, or codes, covenants, and restrictions on solar for your building or neighborhood. You might encounter homeowners' association rules against solar devices viewable from public areas, historic district covenants regarding aesthetics, or other zoning issues such as building height concerns. As mentioned in the "Solar Access Laws" sidebar, in the past few years we've seen many new laws restricting

SOLAR ACCESS LAWS

In the later days of ancient Rome, a combination of enlightenment about the tremendous power of the sun and energy constraints resulting from overharvesting of nearby firewood led to the enactment of solar access laws. Many homes throughout the Roman Empire were heated by the *heliocaminus*, or "solar furnace"—a south-facing room that stayed warm in the winter. As the population increased and the built environment expanded, these rooms became shaded, leading to a ruling in the second century A.D. that a home's access to sunlight could not be violated by one's neighbors. This ruling was eventually incorporated into the Justinian Code of Law.

Likewise, English common law—upon which both United States and Canadian law is based—provided for solar access through the law of ancient lights. Essentially, this says that if a building has had access to the sun for twenty years, then no structure may be erected to shade it. It is also likely that this arose because of the incorporation of solar heating and lighting into English homes, and the resource limits of depleted firewood supplies. As of this writing, there are no known cases of the law of ancient lights being enforced in North America.

We believe we are at the cusp of a similar dynamic, where fossil energy's limits (both its availability and its intolerable pollution), combined with an awareness of solar energy's amazing potential, is leading to the adoption of laws that not only prohibit ordinances that stymie solar installations, but go one step further in guaranteeing a property's access to the sun. Over time, such laws will allow us to rework our towns and cities to be powered by the sun, safe from the limits of fossil fuels and the planetary destruction they produce.

antisolar covenants—and more willingness on the part of homeowners' associations and historic preservation groups to facilitate solar installations. If you have a restrictive covenant or code on your site, will it be worth the effort to fight it? Don't forget the covenant could be outdated or written by well-meaning but uninformed committees in the past, and you could be setting a meaningful precedent by thoughtfully challenging it. It is also entirely possible that the covenant or restriction could be unenforceable because of new solar access laws. Sometimes homeowners' associations will allow variances to their covenants if adjoining neighbors have no objection.

The Condition of Your Roof

One important thing to consider before installing solar collectors of any variety onto your roof is the age and relative condition of your roofing material. Whatever type of solar installation you're considering, it's likely to last at least two or three decades, so it makes sense to have a relatively new roof before proceeding. If your roof doesn't have about ten years of useful life in it, you should replace at minimum the section that will be directly underneath the solar device.

What If My Solar Window Stinks?

In our opinion, there's nothing like adding some renewable energy supplies to the world right on top of everyone's home. But there's no doubt that beyond cost there are a wide range of factors that might make the idea of putting a solar system on your roof or in your yard a hard sell, or quite simply a bad idea all around. The first is not having an appropriate amount of sunlight. You can't live off of solar energy in the shade! A responsible solar installer should caution you against a system if you have poor access to the sun, but an excess of solar zeal might prevent this discussion. It's worth being diligent and checking yourself, probably before you make that first contact.

An important thing to note about solar electricity, as with all electricity, is that it is highly portable. Electricity is already produced far away from where it is consumed, with today's large power plants often sending buzzing electrons hundreds of miles away. Solar electricity works well with **distributed power generation**, or having lots of small producers all giving to and taking from the grid. But solar electricity does benefit some from economies of scale, so large solar power plants can make economic sense.

If you don't have a good spot for a solar system, the portability of electricity means that you can still support renewable energy through green-power or green-tag programs. These are programs that offer homes and businesses the opportunity to purchase a portion or all of their electricity from renewable resources for a small premium, offsetting fossil-fuel-powered generation. The renewable electricity is from a mix of resources like solar, wind, and biomass. While as a homeowner the actual electrons flowing into your home might still come from the coal plant, you can rest assured that the money you are spending to buy electricity is supporting renewable power generation, perhaps a large solar power plant hundreds of miles away. To check if your area is served by one of these providers, check the EPA's Green Power locator (see "Resources").

Understanding Your Current Energy Usage

Providing enough solar energy—whether it's for hot water, electricity, heat, or all three—isn't possible if you don't know how much energy you're currently using. To make sure there's sufficient supply, you must understand your current demand. Of course, there are lots of ways to curb demand, but that's beyond the scope of this book (for lots more on this topic, check out our last book, *The Carbon-Free Home*). We're assuming you've done at least the basics to make sure your home is well-insulated and doesn't consume more than its fair share of electricity and other fuels. This is a vital point—especially from an efficiency, and therefore payback, vantage point. The low-hanging fruit of energy efficiency and conservation (such as properly sealing and insulating your home and switching out incandescent light bulbs for compact fluorescents) reap immediate rewards and often pay for themselves in a few years or even months. While solar energy additions will pay for themselves, assuming proper installation, the payback is often substantially slower—more in the five- to fifteen-year time frame. From a financial perspective, this can be the difference between earning a large rate of return on your investment from conservation and efficiency (7 to 20 percent) to a more modest return on solar energy improvements (0 to 7 percent). We go into more detail about payback and financial implications in the next chapter, "What's Appropriate for Your Budget."

What Types of Energy Does Your Home Use?

Every home uses electricity. Even if it's just a battery for a flashlight, you would be hard-pressed to find anyone (outside of a very small number of orthodox Amish) who shun this most amazing of fuels. Sometimes electricity is the *only* fuel that powers a home, because it can do everything, although not necessarily in the most efficient way. Not only does it run our appliances, computers, stereos, and other entertainment contraptions, it powers our air conditioners, heats up our water, and often heats our homes. Even if the heat is provided by other fuels such as natural gas or propane, electricity runs the fans or pumps that distribute this heat throughout our homes. It even has the potential to run our cars. Because it's used for such a wide diversity of activities, some of which are seasonal, electric bills vary tremendously from home to home and fluctuate throughout the year.

What other fuels power your home? If you're getting deliveries of propane or oil you'll obviously know about it. Natural gas is very commonly available via pipelines in urban and suburban communities. If you haven't already begun, it's time to start accumulating your bills for each fuel type in a folder so you can see how your energy usage varies over the course of the year. Many providers make old bills available through their Web sites, and often a summary of past consumption is represented on current bills (or can be requested by email or phone). If you don't know already, determine which major appliances are being run on which fuel—especially your heat, hot water, and stove.

Electrical Consumption

Besides your electric bill coming every month, you also have information on how much juice you're using in the form of the electric

Figure 2.3. Your electric utility meter is a great place to start to understand broad patterns in your home's electricity usage.

meter on the outside of your home. See the sidebar "Reading Your Electric Meter and Understanding Kilowatt-hours" on page 28 to get a fuller understanding of what your meter can tell you about your electrical consumption. You should also check your electric bill or contact your electricity provider to see if you can find the average for homes in your area for comparison. If your consumption exceeds

READING YOUR ELECTRIC METER AND UNDERSTANDING KILOWATT-HOURS

If you want to find out how much electricity you're using currently, a good place to start is with your electric meter. Utility meters are either analog or digital. Digital meters are quite easy to read as your cumulative consumption is in bold digital numbers on the front. You can easily calculate daily, weekly, or monthly usage in kilowatt-hours (kWh) by writing down that number with a date attached, waiting, and then subtracting it from a future number and date. Analog meters are also readable, with usually five dials with numbers from 0 to 9 and a needle or marker; if the marker is between two numbers, write down the lower number (unless it is between 0 and 9, then write down 9). Write down the dial numbers from left to right, and the complete number will correspond with your cumulative kWh usage. Mark this down with the date and come back later.

To understand what this number on your electric meter is saying, it is useful to look at the examples of hot water heaters and PV panels. A normal electric hot water heater will draw 4,500 watts, or 4.5 kilowatts, while it is on. So if it is on for two hours a day, it will consume 2 hours × 4.5 kW = 9 kWh of electricity. On the flip side of the equation, look at a 2,000-watt (or 2kW) PV array. A 2kW PV array, which can be made up of ten 200-watt panels (each panel about three by five feet), produces approximately 2kW of (DC) power when the sun shines on it. So if you have five sunny hours in a day, your array will produce approximately 5 hours × 2 kW = 10 kWh of electricity in a day (the actual AC energy produced will be slightly lower, because of conversion from DC to AC and other environmental factors). Just because the water heater needs 4,500 watts and the PV array produces 2,000 watts doesn't mean the PV array can't supply the water heater with enough energy—but the time element is crucial to understand. In this example the hot water heater draws 4.5 kWh and consumes 9 kWh in a day, while the 2 kW PV array can produce ~10 kWh.

the local average, determine why. You may need to start with basic conservation measures before installing any solar systems to make sure the solar energy systems will be affordable and effective.

A third source of information on electricity consumption is a household meter (such as The Energy Detective™), which displays instantaneous consumption inside the home. These are becoming more popular with environmentally minded folks, and their installation is being mandated in some locations. For individual appliances, especially ones that cycle on and off like a refrigerator or freezer, a plug-in meter like the Kill-a-Watt can provide cumulative data.

Equipped with a few meter readings and some historical data from your bills, along with an awareness of the power needs of your major appliances, you can start to get a sense of how much electricity your home is using and how to provide this with solar energy. Notice we said solar *energy*, and not solar *electricity*. Although undoubtedly solar electricity can be part of the answer, some of this energy can potentially be provided by other solar appliances that are more efficient. In the example in the sidebar, instead of providing the required energy to heat the water with ten 200W PV panels, a solar water heater could be installed for much less cost and space, and accomplish the same goal.

Determining Daily Average Electrical Consumption

For much of the country, households use more electricity in the summer and winter than in the spring and fall. This is generally because of cooling loads in the summer and heating and lighting loads in the winter. People also tend to be inside more when the weather is less than ideal, using appliances like computers and televisions and stereos. Because of this, it is best to use an average of monthly consumption; try to get an estimate of yearly kWh consumption (add up twelve electric bills), and divide that number by twelve for a monthly average. Then take your average monthly kWh consumption and divide by 30.4 for a daily number (or take a high bill and a low bill and use the average, making sure to also average the days of the two months for the denominator). Keep

Bill Date 05/07/2009
Current Charges Past Due After 06/02/2009

Service From: MAR 31 to APR 30 (30 Days)

PREVIOUS BILL AMOUNT	PAYMENTS (-)	NEW CHARGES (+)	ADJUSTMENTS (+ OR -)	AMOUNT DUE (=)
$19.77	$19.77	$12.64	$0.00	$12.64

METER NUMBER	METER READINGS: PREVIOUS	PRESENT	MULTI-PLIER	TOTAL USAGE	RATE SCHEDULE DESCRIPTION	AMOUNT
191548	414	486	1	72 KWH	RS - Residential Service	13.42
					Supplemental Basic Facility	3.75
	100	133	1	33 KWH	Gen-On Peak Credit-Winter	-1.86
	145	216	1	71 KWH	Gen-Off Peak Credit-Winter	-3.04
					Sales Tax	.37
					Amount Due	12.64

Electricity Usage	This Month	Last Year
Total KWH	72	N/A
Days	30	N/A
AVG KWH per Day	2	N/A
AVG Cost per Day	$0.45	N/A

Our records indicate your telephone number is ???-???-???? . If this is incorrect, please follow the instructions on the back of the bill.

A late payment charge of 1.0 % will be added to any past due utility balance not paid within 25 days of the bill date.

Figure 2.4. Electric bills keep you informed of your energy consumption. Keep in mind that there is often a lot of variation throughout the year, so pay attention to annual consumption as well.

in mind that your current electrical consumption could change dramatically if more people (such as children) or major new appliances (such as a high-definition television or electric vehicle) will be added to your home in the near future.

If electricity provides the hot water and heat for your home, you might want to consider a solar hot water heater or some type of solar heating option before considering a PV system. Once you've reduced your electric loads by providing your major heating needs with these other systems, you'll find that installing the remainder of your electrical capacity with PV panels will be much more affordable. Using electricity for heat is generally very inefficient—especially if that electricity comes from coal-fired power plants, which the majority of our electricity does. By the time the coal is burned far away at the power plant, the steam is heated up, the spinning turbines generate electricity, and then that electricity is transmitted over the utility lines, typically only about one-quarter of the original heat that the coal produced gets turned into heat in your home.

It's also important to note that electricity is highly transmittable over long distances, while hot water and heat are not. If your solar window restricts the amount of space available for solar systems, you may want to consider outsourcing your renewable electricity purchases through a green-power program.

Determining Natural Gas, Propane, and Oil Usage

The other fossil fuels that your home might use are natural gas, propane, and oil. Wood is another fuel (potentially renewable if used judiciously) that might provide significant quantities of energy. Figuring out the exact amount of energy you consume from these fuels is not entirely necessary, as long as you understand which fuel is performing what function in your home. Unlike electrical demand, the services these fuels are providing, most notably hot water and heating, can be evaluated using simpler metrics—like the number of people in the household for solar hot water; and home size, insolation, and heating degree days for solar heating. Each of these is discussed in depth in the solar water heating and solar heating chapters. It is also helpful to know the operating efficiency of your furnace/boiler/woodstove and your hot water heater.

Natural gas, propane, oil, and wood are combustion fuels, meaning they function in your home by being burned and then heating something up, typically the air inside your home, your water, and your food. The last of these is usually a very small slice of the pie (unless you're running a catering business out of your home), and we can ignore it for now (although you'll want to read about solar ovens and cookers in Chapter 8, "Everything Else Under the Sun").

After determining how these fuels function in your home, take a look at your bills (which might be monthly, bimonthly, or seasonal) and see how much fossil energy you're using. Table 2.2 lays out the relationship between all the major sources of energy so that you may do any cross-calculations necessary. The main points to take away from this exercise are the total amount of energy your household is consuming, and the breakdown of how much each fuel (including electricity) provides of the total.

Table 2.2: Fuel Comparisons.
All fuels are unique, and using solar energy effectively has its own challenges. Rather than being a fuel that can be physically stored easily, solar energy is dispersed and accumulates over time. Nevertheless, when properly installed, solar harvesting equipment of all types has the potential to replace much of our fossil fuel use.

Fuel/Energy Measurement (kilowatt hours)	BTUs	kWh
Natural Gas (CCF)	102,700	30
Natural Gas (therm)	100,000	29
Propane (gallon)	92,000	27
Heating Oil (gallon)	130,500	38
Seasoned Firewood (mixed hardwood, cord)	20,000,000	5,860
Solar Energy (hr/m^2)	3,400	1

Understanding total home energy use is a wonderful tool for determining how to proceed with solar additions to your home. As with any challenge, identifying and quantifying usage is the first major step toward finding solutions. Once you've collected all those bills and understand how much energy your household is using, it will be much easier to determine the annual cost of those fuels.

Resources

PV Watts Calculator: http://rredc.nrel.gov/solar/calculators/PVWATTS/version1/

DSIRE solar access laws are on a map with state summaries (you must go to each state through the home page to get details on the laws): http://www.dsireusa.org/userfiles/image/summarymaps/solaraccessmap.gif

EPA Green-Power Locator: http://www.epa.gov/greenpower/pubs/gplocator.htm

CHAPTER 3

What's Appropriate for Your Budget

Energy as a Service

Buying solar energy is an investment—in clean, renewable, nonpolluting energy—but it's also a monetary investment. So it is great news that through 2016 U.S. taxpayers who purchase solar electric and solar hot water systems can take a federal tax credit of 30 percent of the system's installed cost. This is a tax credit, not a tax deduction, so it reduces taxes dollar for dollar. There are myriad other ways that municipalities, state governments, and utilities are making solar more affordable; we'll take a look at some of these as well.

To start a discussion on the economics of buying solar systems, it helps to think first about what it is you are really receiving when you buy any kind of energy. Is it electrons pushed from the power plant or gas from the pipeline, or is it the services energy provides like hot water and lights? If you're thinking about going solar for your hot water and lights and other electrical needs, you are mainly considering substituting one type of service (renewable energy) for another that is readily available (fossil-fueled energy), and it is only natural to wonder what the logic is and what the benefits are.

Think about when you buy a totally different kind of service, let's say a night at a hotel room. You obviously don't wonder about the payback time for your investment in a night's sleep. But you might shop around and decide to pay sixty-five dollars for a hotel room instead of thirty-five dollars—because you know you'll likely get cleaner sheets and a safer, quieter place to sleep. We know intuitively that buying the cheapest service isn't always a good idea. Does that mean we instead buy the most expensive service? No, we weigh the pros and cons, and make a logical decision based on a variety of factors.

Unfortunately, the decision-making factors involved in using energy are mostly invisible and nearly impossible to attribute

Q. Is it "worth it" to buy solar?

A. Solar systems, especially photovoltaics, used to be prohibitively expensive—but not anymore. Prices have dropped dramatically in the past few years and quality has improved—just as tax credits, rebates, and financing options have increased. In many states after getting a system installed and taking relevant rebates, tax credits, financing, and other incentives, you'll be saving money—every month, from the very first month.

Q. Shouldn't I wait for the technology to improve? Will a solar device I install now prove obsolescent in a few years?

A. This technology is tried and true and tested, and warrantied for many years to come. There is no good reason to wait! Cost decreases and efficiency increases happen, but slowly and incrementally.

directly to one source or another. Climate disruption, smog, increased asthma—these are things we know are attributable to the emissions resulting from the combustion of fossil fuels. But every time your refrigerator turns on or you run the dishwasher you don't see puffs of carbon dioxide billowing out. So buying a solar system can be analogous to paying a little more for the clean sheets in the more expensive hotel room—but that's awfully hard to prove, much less explain to a skeptical spouse or neighbor. Since carbon dioxide emissions aren't currently included in the cost of the fossil-fueled energy services we buy, we are left to the mercy of federal and state regulation to level the playing field in other ways. Fortunately, there are many incentives, discussed in depth below, that are currently helping solar gain a foothold, and allowing the fiscal investment in solar to present a viable decision apart from the sustainability equivalent of the "cleaner sheets" investment. There are federal tax credits, state tax incentives and rebate programs, utility incentives, and a bewildering array of financing options all joining the solar party.

For example, in several states you can currently contract a solar electric system for your home with no upfront investment, and ongoing savings accumulate from the very first month's electric bill. The downside of the current options is that they vary so widely from place to place—solar is affordable in California and other states, while it remains inaccessibly expensive to most people in areas without the right incentive structures.

This means you are going to have to do some research and investigation into your current situation.

First we'll take a look at system cost before incentives, and then investigate some of the possible methods of financing and reducing the price tag of the several types of solar systems.

The Cost of Photovoltaic Systems

Photovoltaic (PV) installation costs have been dropping since about 2005, when module prices hit their most recent peak. Most installers will quote a price in dollars, and also in dollars per watt. Dollars per watt is just the total installed cost divided by the "nameplate" DC rating of the PV system (meaning the number of modules multiplied by the power rating in watts of each module—a number that comes from a lab test). So if you have ten 200-watt modules (equal to a 2,000 watt or 2 kW PV array) and the estimate or bid is $17,000, the cost per watt is $17,000 ÷ 2,000, or $8.50 per watt. Installed prices for residential systems before rebates have ranged in the past five years from a high of $12 or $13 per watt to a low of around $5 per watt. Keep in mind that there is a large amount of variability in price based on the size of the system and the specifics of the installation. There are some economies of scale, so bigger systems are usually fewer dollars per watt. (One company, One Block Off the Grid [http://1bog.org/], is trying to use these economies of scale to get bulk-buying discounts for large group installations.)

Dollars per watt is a way to compare the price between installers, but it doesn't allow you to compare the cost of this *service* (renewable energy–powered electricity) to the original *service* (fossil fuel–powered electricity). When you buy energy from a fossil fuel power plant, it is priced in cents per kWh, and we can price PV the same way. For example, if you install a 2 kW PV array for a preincentive cost of $16,000 in our hometown of Durham, North Carolina, you would produce about 2,400 kWh per year of electricity on average for the life of the system, which has a warranty for twenty-five years (see Table 5.1 for an explanation on estimating PV energy production). So with a production of 60,000 kWhs over the course of a twenty-five-year life span, divided by the cost of the system, you pay an average cost of 26 cents per kWh (for a more conservative analysis, you might want to add in the replacement cost for an inverter—they usually have only ten-year warranties, but can last much longer). The average cost of electricity in the United States is about 10 cents per kWh, which means that this particular service, at 26 cents, is looking a little more like a Caribbean resort than a roadside motel.

But a federal tax incentive of 30 percent (and in North Carolina a state tax incentive of 35 percent) bring that original installed cost down to about $6,000. So our renewable electricity cost also drops—down to 10 cents per kWh, the current average price for the dirty fossil fuel electricity you're already buying! So, voila, cleaner sheets for the same price! Assuming fossil-fuel-based electricity prices continue to rise, you're in the black before you know it.

Of course, the downside to this picture is that not everyone is keen to pay for twenty-five years worth of electricity up front. That's where creative financing can play a role, and we'll discuss the array of options below.

The Cost of Solar Hot Water Systems

Heating domestic water (used for dishes, showers, and so on) with solar power is a great investment, in both payback terms or in terms of clean energy. The cost of a solar water heating system for a residence ranges from $3,000 to $9,000, and these systems are also eligible for a variety of incentives—including the 30 percent federal tax credit. Solar hot water can also be used in all kinds of commercial buildings and industrial applications—from restaurants to breweries to hospitals to city office buildings. These commercial and industrial applications take a more nuanced approach to system design, and will generally cost more than a residential system; but along with the investment incentives available, commercial-scale solar hot water is also a great investment.

Domestic water heating on average takes about 15 percent of a normal household's energy budget. Solar water heaters can provide anywhere from 50 to 80 percent of the hot water used, depending on seasonal weather and water-usage patterns. So installing a solar water heater will take a big bite out of that 15 percent portion of your energy bill; a family with an electric water heater might have a hot water bill of between $15 and $50 a month, so savings of $150 to nearly $500 a year are possible. Solar water heating panels usually have ten-year warranties, but life expectancies are at least double that.

The Cost of Solar Heating Systems

The cost of solar heating systems is extremely variable, based not only on the range of different systems available and the variability in size based on needs and climate, but also because of a lack of clarity in federal tax law. A solar air heater installed for supplemental heat for one room might cost as little as $1,500, while a passive solar retrofit could run into the tens of thousands.

Solar heating systems—active or passive, radiant or forced air (see Chapter 7, "Solar Heating," for descriptions)—have been somewhat under the radar of legislatures and utilities when it comes to rebates and financing. Unfortunately, when Congress extended the renewable energy tax credits to the end of 2016, they did not change the wording to include solar heating systems for the 30 percent federal tax credit. What is included is the original solar water heating credit—and a line that refers to "other solar electric technology." Stephen has installed several solar air heating systems since the tax credit went into effect, and discussions over whether there is any way to link solar heating options to the federal tax credit have produced a variety of opinions—from both the homeowners and their various tax attorneys.

So can you take the 30 percent federal tax credit or not? Unfortunately the answer is—it's up to you and your tax lawyer's interpretation of the law. Providing heating from solar energy certainly seems to be within the spirit of the law, since the law was passed to encourage the use of solar energy generally (i.e., not just solar electricity). And it would seem the provision for "other solar electric technology" was included because lawmakers knew they were unaware of the myriad choices and wanted to be comprehensive without exhaustively listing all solar technology. If you install a solar air heater using a small fan powered by a PV panel, for instance, does the entire heater qualify as "other solar electric technology," or just the PV portion, or just the PV module and fan? Like much of tax law, the answer is—who knows? Some homeowners will feel comfortable taking the credit, others won't. At a minimum, using a Solar Rating and Certification Corporation (SRCC)-rated heater would seem to be a solid tax move (and good idea in general), since SRCC certification is required for taking the solar water heating credit. If you've got a spare minute or two, it might not hurt to send your representative an email asking them to amend the law to specifically include solar heating.

A 30 percent reduction in price is substantial, and this will certainly change the economics of how easy it is for you to afford the purchase of solar heating equipment. From an efficiency viewpoint, all else being equal, solar heating would likely fall in between solar water heating (at about 60 percent efficiency) and solar electricity (about 15 percent) in terms of energy payback, since solar heating is a thermal technology similar to solar water heating, but only used about half the year instead of year-round (60 percent × 50 percent = 30 percent efficiency).

All else is not equal, however, since solar heating has generally gone unnoticed not just by federal legislatures but by state legislatures as well. That is not the case for us here in North Carolina, where solar heating explicitly shares the same tax benefit as solar water heating and solar electricity. Determining your particular locality's attitude toward solar heating will require some research at the Database of State Incentives for Renewables and Efficiency (DSIRE) Web site (http://www.dsireusa.org/solar), which will also list any local and utility incentives.

One method of using solar heating technology that should be more likely to qualify for the federal tax credit is to incorporate a radiant heating system into a solar water heater installation. Since there is not a limit to the dollar amount of this tax credit, expanding the solar water heating system to include solar radiant heating seems like it would be covered by the credit. Unfortunately, since it doesn't specifically include solar water heating used for radiant heat, you'll have to consult a tax attorney.

Other Solar Device Costs

Solar pool heaters are another great investment in solar if you have a pool you are currently heating with gas or electricity, which can become extremely expensive. Cost is variable, but it is generally less than $5,000. The federal tax credit does not apply, but many states do have extra incentives (again, check the DSIRE Web site). If you are spending $500 a month to heat your pool in the shoulder seasons (not unusual in colder climes), it wouldn't take long for your solar pool heater to pay for itself!

Help Paying for Your System

In the solar world, there are production-based incentives and there are installation incentives. Production-based incentives—like feed-in tariffs (FITs), green-tags, renewable energy credits (RECs), or alternative energy credits (AECs)—help pay for a system slowly over the energy-producing life of the system by cash or credit payments per kWh or BTU of renewable energy produced from a solar system. Installation incentives—direct cash payments via state or utility rebates, tax credits, tax deductions, or sales tax exemptions—

Table 3.1. How to Pay for Your System:
Direct Cash Incentives, Production Based Incentives, and Financing

Possibilities for Reducing the Overall Upfront Cost	Financing Possibilities
• Tax credits (federal, state, or municipal) • Tax deductions and accelerated depreciation (federal, state, or municipal) • Sales tax or property tax exemptions • Direct cash payments or installation rebates (municipality, state, or utility) • Production based incentives (feed-in tariff [FIT], green-tags, renewable energy credit [REC], alternative energy credit [AEC], renewable premium payment)	• Power purchase model: buy production per kWh or BTU, but do not buy system components or installation • Leased system: lease system for a set monthly fee and use energy generated • Energy efficient mortgage • Green loan • USDA REAP loan for farmers and rural small businesses • Home equity loan or line of credit • Credit card • Property Assessed Clean Energy Loan (PACE)

help pay for a percentage of the system when it is first installed, and are based on the cost or size of the system. In addition, the various methods of financing can take the initial cost of a system down all the way to nearly nothing. Depending upon where you live or own property, you will have no, one, or many layers of these tools available to help you (for a real-world example, see the sidebar on the State of Pennsylvania, which has both production and installation incentive policies). Check out Figure 3.1 and Figure 3.2 to find out what's available in your state.

The nice thing about installation incentives is they almost immediately reduce your expenditure (at least within the tax year). This is especially true if your installer will front the cost of the installation rebate, something that often happens in states with established solar industries like California and New Jersey. Production-based incentives, on the other hand, can take a while to pay out, but are based on actual system production numbers—so they incentivize you to demand a high-quality installation, as you will only get paid based on the renewable energy you produce!

Rebates can come from a variety of sources—states, municipalities, and utilities are the most common. California has had a long

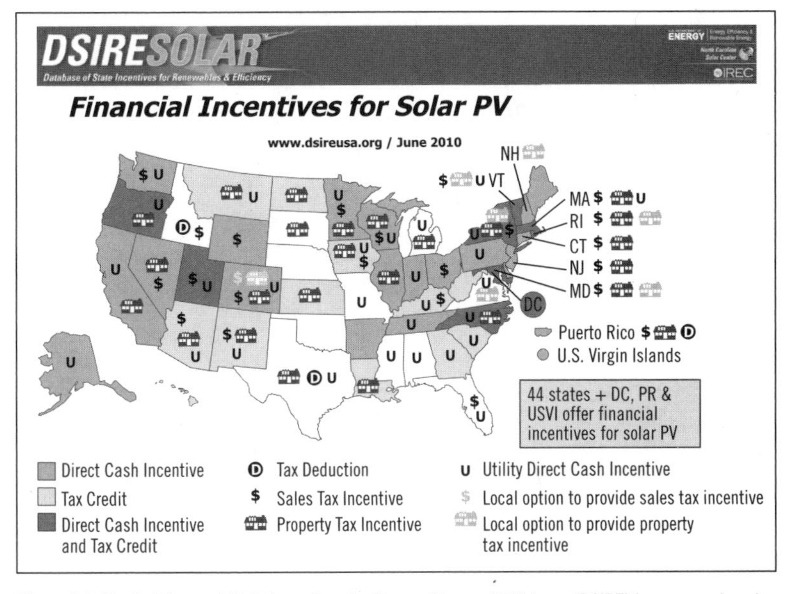

Figure 3.1. The Database of State Incentives for Renewables and Efficiency (DSIRE) is a comprehensive source of information on state, local, utility, and federal incentives and policies that promote renewable energy and efficiency. Established in 1995 and funded by the U.S. Department of Energy, DSIRE is an ongoing project of the North Carolina Solar Center and the Interstate Renewable Energy Council. Be sure to check the DSIRE Web site for the latest information. SOURCE: US DOE/NC SOLAR CENTER/IREC.

running rebate program, which is why approximately 80 percent of the solar in the United States has been installed in California. Depending upon who is offering the direct cash rebate, it might either be a fixed percentage of the system cost with a cap (30 percent of the system cost with a $4,000 cap, for example) or a sliding scale based on the wattage of the system (for example, $2/watt regardless of system size). Many direct cash-rebate programs ramp down over time, with the per-watt or cap amount decreasing as more systems are installed in an area. Often the rebate, especially if a utility is offering it, comes with strings attached—all the green-tags or RECs the system creates might be exchanged for the rebate. Read the fine print and ask your installer plenty of questions!

Of the scores of solar systems Rebekah has installed, many were financed in some way. Home equity lines of credit; second mortgages; green, USDA, or small business loans; community incentives—there is a long list of possibilities. If the financing for a solar system costs less

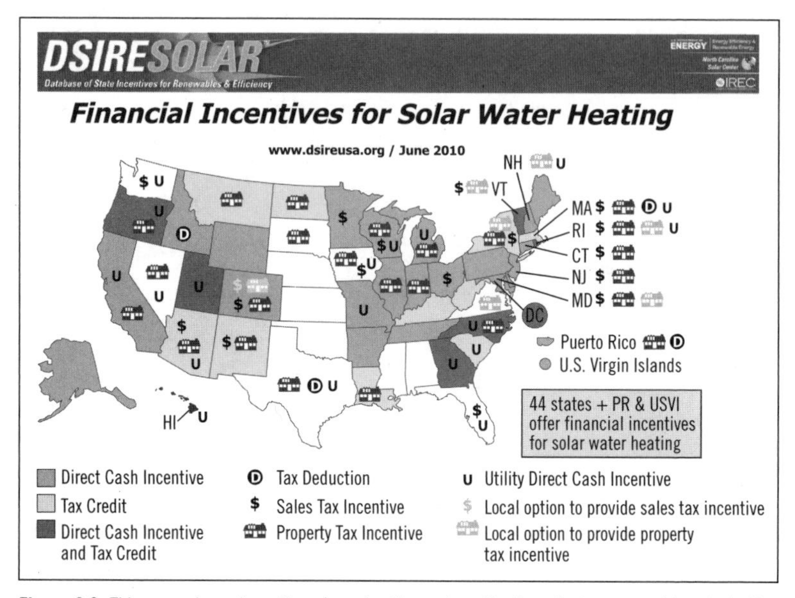

Figure 3.2. This map shows incentives for solar thermal applications (hot water and heating). Also included at the main DSIRE Web site is information on tax credits for efficiency, green loans, and regulations related to green building. SOURCE: US DOE/NC SOLAR CENTER/IREC.

per month than the solar system will save you on energy bills, then it is an easy financial argument to make. Banks, mortgage companies, and the federal government lend money to finance solar systems, but there are sometimes other options that are easier. SolarCity, an installation company that started in California, finances the PV systems it installs. SolarCity's leasing program means homeowners can pay a fixed cost per month for their PV system, and use the electricity generated—but don't buy or maintain the system, which translates into little money upfront. This is similar to what's known as a **power purchase agreement (PPA)**, meaning the building or property that hosts the system doesn't own the system, but only purchases the energy created while someone else owns and maintains the hardware. The price paid for solar electricity is on par with or cheaper than fossil-fueled electricity. SunRun is another company offering residential leasing and PPAs in states with established solar incentives. Some states themselves are even getting into the action—Connecticut has a homeowner solar energy leasing plan as well!

The Federal Housing Authority, Fannie Mae and Freddie Mac, and many private banks offer energy efficient mortgages; in conjunction with an energy audit that rates the energy efficiency of the home (often the Home Energy Rating System (HERS) is used), these mortgages can be used to roll the cost of a solar system into a new home purchase or refinance. The reason an energy rating is necessary is that an energy efficient mortgage may only be available if the solar system proves cost effective—meaning it will reduce monthly bills more than the added monthly cost of the system. For farms, ranches, and rural businesses, there is a USDA program—the Rural Energy for America Program (REAP) guaranteed loan program.

Another rather new way to finance systems is through what are called Property Assessed Clean Energy (PACE) bonds. This type of bond can be issued by a municipality or local government (the idea originated in Berkeley, California). Property owners get a loan from bond proceeds to finance a renewable energy system or other energy efficiency measures, and then repay the loan via an annual assessment on their property tax bill (or another locally collected tax bill specific to the property). Usually PACE financing must be authorized by legislative action to issue assessments on a specific customer's taxes, but the local government budget is not affected because all administrative costs are covered by the property owners who receive the loan. If the property owner moves and sells the PACE-affected property, the repayment obligation and renewable energy system transfer with the property. Obviously not every municipality currently offers PACE bonds, but a list is maintained by the DSIRE Web site (see "Resources").

What the Heck's a REC or a FIT?

REC is an acronym for renewable energy credit—also known as an AEC (alternative energy credit), SREC (the *S* is for *solar*), or green-tag. RECs exist as a way to buy and sell the "green" value of energy created through renewable resources (that value being related to the carbon emissions that were avoided by using green power instead of fossil-fueled power). The standard definition of a REC is one megawatt hour (MWh) of energy produced from a renewable resource. The actual electrical energy created is sold for a different and separate value, in cents per kWh. What precisely qualifies as a renewable resource is the subject of much debate, but fortunately solar is high above that debate—rest assured solar is a green renewable resource.

But what qualifies for a REC is just the tip of the iceberg when it comes to the murky confusion surrounding this issue. Where, how, what, and who buys and sells RECs varies throughout the country and throughout the world. What is important to you as a homeowner or business is that if you can find a buyer for your RECs, you just stepped on the accelerator pedal for the speed of payback from your solar investment.

The leading region in the U.S. for developing a REC market has been the northeast zone of the country. New Jersey has a constrained internal market, but there are also multistate trading platforms, private aggregator businesses who buy RECs directly from homeowners, and many other states developing trading platforms. RECs (remember, 1 MWh) have gone for more than $700 apiece and for as little as $10 or $20. RECs are a tradable commodity, so the price is set on the open market. That doesn't mean that you can't lock in a price—some buyers are willing to sign five-, ten-, fifteen-, or even twenty-year contracts at a set price, and some buyers even offer an upfront payment option with a one-time payment for five years of future production. Whether that set price will be above or below market value in the future is a gamble the seller and buyer both take.

As a solar electricity producer, you'll create around one REC a year per kW of installed PV, perhaps a bit more depending on the solar window of the site. Most homeowners don't produce a large quantity of RECs compared to utility-scale solar installations, so they end up selling them to aggregators who take a little percentage of the price and sell bigger bundles of RECs. The usual suspect for the end-purchaser of RECs is a utility trying to meet a legislated renewable energy portfolio standard that determines a percentage of electricity that needs to come from renewable resources. Another type of buyer might be a corporation like Whole Foods or WalMart that wants to certify for PR reasons that a certain percentage of the electricity they use is "green." You can in fact sell your RECs to anyone willing to buy them, even if that means the gas station down the street from your house.

RECs can be sold from solar thermal systems as well as PV, but it takes a little more effort to track the BTUs a system produces, and eligibility varies from place to place. To learn more about REC prices, read the sidebar about the state of solar in Pennsylvania.

There is one more solar acronym headed our way! A feed-in tariff, feed-in rate, or FIT as they are becoming widely known, is a supplemental payment over and above the cost of electricity to a PV power producer, paid per kWh. Unlike RECs, these aren't tradable commodities, but rather act in just the reverse way of an electric bill. A utility, municipality, or governing body decides to promote renewable energy by paying a premium for renewable electricity generated. A contract is devised where for a fixed term of years (usually five to twenty), a set price (that may rise with inflation) is paid for the renewable energy a home or business owner generates. For example, through the Gainesville Regional Utility the city of Gainesville, Florida, has set up a FIT, with contracts for systems signed in 2009 paying 32 cents per kWh for twenty years. The program has a cap on the installed wattage allowed into the program, and applications filled up the 4 MW initial year allocation queue immediately when the program was announced in 2009.

The FIT is paid for by rate payers in the utility's service area, who each pay a very small amount more on their bill to pay for the clean energy supplied by producers. In Gainesville, implementing the FIT has added about 70 cents per month onto electric bills, but a survey of the area shows 75 percent approval ratings for the FIT program.

Another program very similar to Gainesville's is found in Tennessee. The Tennessee Valley Authority oversees Green Power Switch, which pays a premium of 12 cents over to the normal cost of electricity per kWh for solar generation. For example, if you pay 10 cents to buy electricity per kWh, you would receive 22 cents/kWh for your solar-generated energy. This isn't strictly speaking a FIT, but has the same advantages.

The state of Vermont and Ontario, Canada, also have FIT programs in place that have gotten off to roaring starts, with companies and homeowners racing to apply. The best known FIT originated in Germany, where the popularity of the program made Germany the world leader in installed photovoltaic systems for much of the last decade.

The Federal Tax Credit

Please check with your tax accountant and attorney before making any assumptions regarding solar renewable energy system tax credits. The most exciting recent news in the U.S. solar industry is the American Recovery and Reinvestment Act of 2009, which gives an extension of the 30 percent renewable-energy photovoltaic and solar thermal consumer tax credit for systems placed in service before December 31, 2016. This credit does not apply to solar pool heaters, but performance-rated solar water heating systems do qualify if they supply at least half the energy used for a dwelling's hot water.

In other good news, the extension of this 30 percent tax credit also lifted a $2,000 cap that had previously limited the credit for residential systems. For systems placed on homes (although the home doesn't have to be a primary residence), this 30 percent is not a grant or tax deduction but rather a tax credit that is taken in the year the system goes into use. While this is a great incentive, it is important to realize that you must pay the installer this 30 percent up-front—and you won't receive the benefit of the tax credit until you file the year's taxes. Commercial building owners are also eligible for the 30 percent tax credit (plus accelerated depreciation of the system cost).

The 30 percent credit includes all "qualified expenditures," which means materials, labor, wiring, and so on. If the credit

THE STATE OF SOLAR IN PENNSYLVANIA: AN EXAMPLE OF MULTIPLE INCENTIVES IN ACTION

Pennsylvania offers both installation incentives and production-based incentives.

Since 2009, the Pennsylvania Sunshine Solar Program provides a direct cash installation incentive in the form of a rebate for solar installations. The rebate amounts differ for photovoltaic and solar thermal systems, and also decrease over time as more systems are installed (this is a common feature of rebate programs good for early adopters!). There is a laundry list of rules and regulations attached to the rebate concerning qualified installers, qualified installation sites, and so on (see "Resources" for further information). Table 3.2 focuses on the Pennsylvania Sunshine rebate for solar thermal and solar water heating (which cannot be used for pool or hot tub heating). The rebate is capped at $2,000 or a fixed percentage of the system cost, whichever is less; Table 3.2 shows how rebate percentage amounts are lowered once a certain plateau of installed systems is reached.

Pennsylvania also has an alternative energy portfolio standard, which means that production-based incentives exist as well, and renewable energy producers can sell what are called alternative energy credits (AECs). AECs are exactly like RECs—tradable certificates representing clean energy benefits, issued to certified producers for each megawatt-hour of renewable energy (remember this is a payment separate from payment for electricity generation). AECs have been trading recently in the $200 to $700 range, or 20 to 70 cents a kilowatt-hour, a great incentive!

exceeds taxes due, the extra can be rolled over year to year at least until 2016 (after which time the laws are unspecific). Note also that the maximum credit is $2,000 for systems placed in service before January 1, 2009. The Solar Energy Industries Association (SEIA) has published a three-page document that provides answers to frequently asked questions regarding the federal tax credits for solar energy; see "Resources" at the end of the chapter.

Table 3.2. Residential and Small Business Solar Thermal Rebates, Pennsylvania Sunshine Solar Program (Maximum Rebate Capped at $2,000).		
Step Number	Cap on Number of Solar Thermal Systems in Step	Rebate Amount (Percentage of System Cost [Maximum $2,000], Decreases with Increasing Participation)
1	1,500 systems	25 percent
2	1,500 systems	20 percent
3	1,500 systems	15 percent
4	1,500 systems	10 percent

To summarize, there is a somewhat bewildering spectrum of methodologies being used across towns, cities, states, and indeed the world to incentivize solar energy. These financial incentives are not misplaced, for they increase clean energy and reduce airborne pollutants, including those that cause asthma and climate disruption; support distributed generation as a way to keep energy sources close to home; and help solve the problem of an aging electricity supply, transmission, and distribution infrastructure. These are serious issues facing our civilization and affecting our future prosperity, and renewable energy is one tool in our toolbox. Until the external effects of fossil-fueled power are internalized in the price, renewables will likely need crutches like those we've discussed to play on the same field.

Resources

DSIRE Solar Financial Info: http://www.dsireusa.org/solar/solarpolicyguide/?id=26

http://www.dsireusa.org/incentives/index.cfm?EE=1&RE=1&SPV=0&ST=0&searchtype=PTFAuth&sh=1

DSIRE State Policy Comparisons: http://www.dsireusa.org/solar/comparisontables/

DSIRE Solar Specific Policy Information: http://www.dsireusa.org/solar/

Solar Energy Industry Association (SEIA) Tax FAQ: http://seia.org/galleries/pdf/SEIATaxManual_v3-0_FAQ.pdf

Average Retail Electricity Prices from the U.S. Energy Information Administration: http://www.eia.doe.gov/fuel electric.html

General Information on RECs from the U.S. Department of Energy: http://apps3.eere.energy.gov/greenpower/markets/certificates.shtml?page=0

U.S. Department of Housing and Urban Development Guide to Energy Efficient Mortgages: http://www.hud.gov/offices/hsg/sfh/eem/eemhog96.cfm

USDA Rural Energy for America Program Guaranteed Loan Program (REAP loan): http://www.rurdev.usda.gov/rbs/busp/9006loan.htm

PPAs, Financing, Lease Programs, or Reduced Cost Programs

SolarCity: http://www.solarcity.com/

SunRun: http://www.sunrunhome.com/

One Block Off the Grid: http://1bog.org/

State of Connecticut: http://www.ctsolarlease.com/index.php

Incentive Programs

Gainesville Regional Utility Feed-In Tariff Program: http://www.gru.com/OurCommunity/Environment/GreenEnergy/solar.jsp

Tennessee Green Power Switch: http://www.tva.com/greenpowerswitch/partners/index.htm

Vermont SPEED Feed-In Tariff Program: http://vermontspeed.com/standard-offer-program/

Pennsylvania Sunshine Program: http://www.portal.state.pa.us/portal/server.pt/community/in_the_news/10475/pa_sunshine_solar_program/553019

Pennsylvania AEC Information: http://paaeps.com/credit/

New Jersey SREC Information: http://www.njcleanenergy.com/renewable-energy/programs/solar-renewable-energy-certificates-srec/new-jersey-solar-renewable-energy

CHAPTER 4

Getting Ready for the Installation

Once you've decided to take the solar plunge, the hard work begins—with the first step of finding and interviewing trustworthy installation companies. Something to remember is that a majority of solar installation companies across the country handle both solar thermal and PV installations. This doesn't always hold true, and some companies might be dabblers rather than experts in one or the other. But if you are planning to have both PV and solar thermal installed, you can have both site visits done simultaneously.

STEP 1: Create a List of Qualified Contractors

Caveat emptor—generally speaking any or all of the contractor lists mentioned below are just a first stop in the vetting process, and don't mean anything concrete except that those contractors met a bare minimum of requirements, perhaps having a contractor's license or passing an entry-level exam. Also beware that any list of contractors you find on the Internet might include not just regional but national—or even international—installers that have little experience in your area.

First and Foremost: Check Your Rebate Programs

States or utilities with rebate programs in place likely have a list of approved installers—ones that have met basic accreditation—that you *must* choose from to qualify for the rebate. So first check http://www.dsireusa.org (Figure 3.1 and Figure 3.2) for rebate or incentive programs in your area, and then follow up with each program manager or the program Web site.

For example, Pennsylvania maintains a list of qualified installers for their Pennsylvania Sunshine rebate program, and the Sacramento Municipal Utility District (SMUD) maintains a similar

Q. How do I know if a solar installer has the right knowledge?

A. Look for state contractor's licensing, and certification and training from the North American Board of Certified Energy Practitioners (NABCEP) and the Interstate Renewable Energy Council/Institute for Sustainable Power Quality (IREC/ISPQ). NABCEP is a widely recognized certification procedure for solar thermal and PV installers. Beware that NABCEP certificates comes in two stages for PV—an entry-level exam, and an installer certificate. ISPQ certifies solar teaching institutions, curriculum, and instructors.

Q. How do I make sure I'm getting all the rebates and tax credits possible?

A. Check the Web site http://www.dsireusa.org/solar and find your area.

Q. How do I know if I'm getting a fair price or being overcharged?

A. Get multiple bids, and do your homework!

list for their programs (see "Resources" at the end of this chapter). If there is no list maintained, ask the rebate or incentive program manager for other installer requirements necessary to access the incentive, such as the North American Board of Certified Energy Practitioners (NABCEP) certification or a contractor's license, and make that the first bar for the installers you choose to interview. *There may be more than one incentive available in your area, and you must make sure your contractor meets requirements for all programs.* After clearing this hurdle, you will find a number of ways you can learn about possible system installers.

Online Installer Reference Sites A growing community exists of online sites that offer installer referral services. These sites are all free to users, and commonly are funded via installer fees or grants. They vary in quality, but many—like the nonprofit solar-estimator.org—include real customer references, which can be not only entertaining but sometimes enlightening to read. These sites are also a good place to pick up general solar knowledge and lingo.

Another type of installer reference site is maintained by NABCEP of their certified PV installers. NABCEP develops national voluntary standards and certifications for renewable energy professionals. The NABCEP program is very thorough, and has created a high national bar for exemplary solar professional accreditation. See "Resources" at the end of this chapter for installer reference Web sites.

Good Old Word of Mouth. People with solar on their house or business love to talk about it. Find out if there is an annual American Solar Energy Society (ASES) solar tour in your area by checking their Web site (see "Resources") and visit some local systems; you can even just stop and knock when you see a system.

Listing Agencies. Agencies like the California Energy Commission, local universities, environmental centers, or solar centers often maintain lists of solar installers (see "Resources"). As an example, if you live in Oregon, you can visit the Oregon Department of Energy solar program Web site to find out about incentives, events, and contractors in Oregon. You could also visit the Oregon chapter of the Solar Energy Industries Association (OSEIA), which maintains a contractor list as well.

STEP 2: Thoroughly Vetting the Contractors on Your List

While some parts of the country might have only two or three installers to choose from, places like New Jersey have hundreds. So how do you choose from a list of sixty qualified installers? It's going to take some legwork! But think about buying a car—you check *Consumer Reports*, you read online reviews, you stroll around the car lot. This system should last a heck of a lot longer than your car, and might cost more to boot! So take some time with this, and you'll get a satisfactory result.

Web Sites. Spend a few hours doing Internet research on potential candidates and visiting Web sites. On the Web site are there pictures of recent projects? A list of qualified employees? Company background information? Does the company *frequently* work in your area? Are they a licensed contractor? Some companies fill up their sites with solar education links or pictures from projects they haven't done. Put on your skeptic's glasses and read through the blinking lights.

Check Licensing. Every jurisdiction is different, so make sure that the installers you choose to interview have their licensing in order for your inspections, plus state, utility, or other rebate requirements. Many installers post copies of these documents on their Web sites. Some states have a solar-specific contractor's license, like the "C 46 Solar Contractor" in California. Other states mandate an electrical contractor's license. Call your state's contractor licensing board and ask what is necessary for a solar installation.

Consumer Protection Office. If you have any concerns about the validity of a business, cross-check with your state consumer protection office (see "Resources").

STEP 3: Setting up the Site Visit

First Call or Email. Most solar companies will have a form inquiry on their Web site for you to fill out with general project information. Can you estimate hours of sun received on average for your spot? Do you know which direction it faces? For a PV project, have your utility bill handy. Basic roof measurements can be extremely useful as well. For a solar thermal project, check the gallon capacity on your tank, and whether it is gas or electric.

Is your inquiry answered quickly? Remember that the best companies are the busiest and not necessarily the largest, so try not to be too impatient—everyone gets overwhelmed from time to time. If you start investigating a project in December, expect to wait as solar companies are at their peak season trying to get projects completed before the end of the tax year.

Don't forget that solar installers get pummeled with inquiries that lead nowhere. Lots of people are just "tire kickers" looking for a little solar education and that's it. But once you make it clear that you are serious, the installer should take you seriously as well.

Figure 4.1. Solar electricity installations have become much more streamlined as the industry has matured. The entire installation often takes only a few days once a contract is signed and the equipment delivered. PHOTO COURTESY OF SOUTHERN ENERGY MANAGEMENT.

If you tell the installer you are getting multiple bids, they should treat you with kid gloves and give you their lowest price first, but they may not be as open with information as they would have been otherwise.

Schedule Site Visits. Once you are in communication with the installers, the next step will be to schedule site visits. A site visit is for the installer to measure the solar window and installation area, inspect the electrical or plumbing systems, and prepare to estimate costs. Some installers will come prepared with a system quote to print after measurements are taken, but other systems are too complicated to quote on site. Don't go overboard, but try to get three quotes if you have access to that many qualified installers.

Installers will in certain circumstances charge a fee ($50 to $150) for site visits. The fee will often be refunded into the system cost if the customer buys a system. It depends on the area, the competition, and how serious the installer thinks you are as a buyer. Site visit fees are intended to discourage tire kickers (thus saving the installer lots of time and driving), not to make money for the installer. So if you tell the installer you've already done your homework, you know what you want, and you know you have enough sun—and you've got the cash to purchase—they could very well waive the site-visit fee.

Questions for the Installer during the Site Visit

- Technology: Does the installer represent only one product line? If so, is that the best choice for me?
- What kind of training have your installers completed?
- Who is going to be performing this work? Will there be a licensed individual on site?
- What warranties are offered on the parts and installation labor?
- Is there a service department? Who will service the system in the future?
- Does the system include data monitoring? If so, is there a yearly ongoing fee?
- Timeline: When does the installer's schedule open up?
- Is the quote a fixed bid or a changeable estimate? What is the process for unexpected costs?
- What deposit is expected and what is the payment schedule?
- If there's a direct cash rebate, does it come as a refund check or get subtracted from the up-front system cost at contract signing?
- Has the installer worked in this area before, and do they know the inspection system?
- What inspections are needed?
- Is the installer going to take full responsibility for ensuring the system passes any inspections that are needed and pay for required permits?
- Will the installer complete all necessary interconnection

paperwork and rebate paperwork? Are there extra fees for the interconnection applications?

- Last but certainly not least, ask for a list of referrals to contact and the closest completed projects to visit and inspect.

Step 4: Referrals

Once you get a list of referrals, spend some time checking up on them. If you have the time, ask to visit a completed project. Ask questions. Did the installer come when promised? Did they clean up the job site? Provide a safe working environment for employees? Exercise caution and use respectful language around children? Deliver the project as expected and at the quoted cost? Any lingering issues?

If you visit a completed project, look for a clean and neat installation, no dangling wires, clear labeling, no error lights, and an informative manual for the homeowner with operating instructions, warranties, and contact information.

Step 5: Timeline

Once you sign on the dotted line, the time to installation could be as little as a few weeks to as long as a year. Generally speaking, installations should be completely finished within a few months. If there is a holdup, it will likely be related to either a backlog in a rebate program, a delay in material delivery, or weather. There have been times when the wait for specific PV modules could stretch on for quite some time. Fall is normally the busy season for solar installers, as systems need to go in before the next year to take advantage of tax credits for that year. Roof-mounted residential PV or hot water systems can usually be installed in one day to one week. Ground- or pole-mounted PV systems take longer, as groundwork, concrete, trenching, and so on take more time. Larger systems might also take a bit longer.

Figure 4.2. Solar water heating panels being installed on a residential roof. Once the placement of the water heater and the panels is finalized, installation is often completed in a day. PHOTO COURTESY OF SOUTHERN ENERGY MANAGEMENT.

Step 6: Paperwork

Interconnection Applications. Especially for PV systems, there can often be pages and pages of interconnection paperwork—first for your electric utility, then state utility commissions, federal regulatory agencies, and rebate programs. Your installer should make this a smooth process and be comfortable with the required paperwork, supplying you with documents to sign and informing you of the progress of your applications. Experienced installers will likely encounter fewer snafus getting past the paperwork hurdles.

Inspections. The inspection process is unique to every jurisdiction. Permit fees evolve over time, and the permitting process evolves as well. Every utility-connected solar electric system must be inspected, and solar thermal systems should be inspected as well.

A well-vetted installation company will know how to conduct your system through the process without any trouble. Ask for a copy of the final approved system inspection for your records.

Insurance. You will need to inform your insurance company of your new solar system in case of vandalism, theft, or storm-related damage. Usually this is a painless process. It is often necessary to provide a copy of insurance for interconnection paperwork.

Step 7: Installation Complete!

Once the system is completely installed and the inspections have been passed, then the last step is to get a complete walk-through of the system and look over the system manual with the installer. The system will be fully commissioned at this point. You should learn what to look for to make sure the system is operating properly and where any error lights or system shutdown and bypass switches are located. If you have a data monitoring system, check that it is operational. Your installation contractor will expect full payment of any remainder at this point. Congratulations, and welcome to the world of solar energy!

Resources

For links to all state rebate and incentive programs to check for qualified installer lists: http://www.dsireusa.org/solar

General Info:

American Solar Energy Society: http://www.ases.org/
Solar Energy Industries Association: http://www.seia.org/

Installer Reference Web sites:

http://www.nabcep.org
http://www.solar-estimate.org
http://www.getsolar.com

http://www.findsolar.com
http://www.coolerplanet.com

Regional Rebate program installer list examples:

Pennsylvania: http://files.dep.state.pa.us/Energy/Energy%20 Independence/EnergyIndPortalFiles/solar/installers/approved_ pv_installer_list.pdf

Sacramento Municipal Utility District: http://www.smud.org/ en/community-environment/solar/Documents/solar-pv retrofit-feb2009.pdf

New York State Energy Research and Development Authority (NYSERDA): http://www.powernaturally.org/ Programs/Solar/Installerspv.asp?i=1

State Agencies, University, Solar Center list examples:

California Energy Commission: http://www.gosolarcalifornia .ca.gov/database/search-new.php

Oregon Department of Energy: http://www.oregon.gov/ ENERGY/RENEW/Solar/index.shtml

Oregon SEIA: http://www.oregonseia.org/contractors

North Carolina Solar Center: http://www.ncsc.ncsu.edu/ gosolarnc/index.php# (under solar contractors)

State Consumer Protection Offices: http://consumeraction .gov/state.shtml

CHAPTER 5

Solar Electric (Photovoltaic) Systems

Electricity is magical. As a constant, quiet, steady supply of energy, it is able to power an amazing variety of gadgets, from the kitchen blender to the office computer to the living room stereo—and possibly even the automobile out in the garage! Until recently, access to this magic was only available from the nearest power plant, which typically generates its electricity by burning gargantuan quantities of planet-choking coal. But a revolution has been budding, one that for the first time in history has the potential to allow us access to the many wonders and conveniences that electricity offers without ruining our collective home (Earth) in the process.

As we fleshed out in the Introduction, solar electricity started out as a means to power satellites in space, then became a way to fuel the lights of starry-eyed homesteaders, and has now become a viable way to power the homes and lives of regular folks in regular towns all across America and the world. A number of factors have converged over the last several decades that have made electricity from the sun into a real and accessible phenomenon. The first is the consistent and steady progress made in every aspect of solar electric generation, from the panels themselves to the racks that hold them, to the inverters and other highfalutin electronics that make the juice usable in the home. The second factor is the ability to make this electricity available not just for the home or office where it is produced, but for any user on the electrical grid.* And the third factor is a genuine commitment by governments on every level to assist in this conversion to sustainable power not only through tax breaks, rebates, and credits, but also through laws that mandate certain percentages of utilities' power capacity be made by renewable means such as solar electricity.

* "The grid" is shorthand for the local supply, transmission, and distribution of electricity, provided and maintained by an investor-owned, cooperative, or municipal utility that is usually interconnected with larger multistate and even international electric grids.

Q. Can I really send solar electricity back to my utility and "spin the meter backward" for credit on my bill?

A. In nearly every state in the United States, and in many other parts of the world, yes. The technology works perfectly, but interconnection paperwork and regulations are still being written in some places. See http://www.dsireusa.org under "Net Metering" to find out about your specific location.

Q. If I install a photovoltaic (PV) system can I have power during blackouts or brownouts, when the grid is down?

A. Yes, but only if you install a more complicated system with battery backup. Grid-direct PV systems (without batteries) won't work when the grid is down.

Q. Do I have to get enough solar panels to make all the electricity I want to use?

A. No, if you already have an electric utility provider, you can make a little or a lot—it depends on your space and your budget. Your electric utility will make sure your lights stay on regardless!

Q. Where I live there is very little sun in the winter, and my roof is often covered in snow. Does it really make sense to install solar equipment up there?

A. Long sunny summers in northern climes or high elevations mean annual average production can be as high as anywhere else in the country (see Table 1.1); Colorado is a good example of a snowy place that's great for solar electricity! A good installer can help site panels to maximize production and help snow shedding.

Q. I've seen those ads to build your own solar electric panels—is that a good idea?

A. There is a definite possibility a homemade panel with poorly soldered connections could start a fire. Building your own panel might be a good science project, but it is absolutely not a good idea to install electrical equipment on your roof that hasn't been tested and listed (known as UL listing; UL is the Underwriters Laboratory and has a lengthy, well-designed testing procedure for PV modules).

Grid-Tied versus Off-Grid

Tying solar electric power into the grid is probably the biggest of these breakthroughs. For decades, residential PV systems were always **stand-alone or off-grid** power systems. In stand-alone systems, power generated by the solar modules is either consumed immediately in the home or directed to batteries for storage and later use. If the batteries are full and there are no electrical loads within the home, the sunshine falling on the PV panels goes unused. The fact that each stand-alone system also has to have its own storage capability in the form of a bank of short-lived batteries greatly increases the cost of both installation and long-term maintenance. Creating a PV system that can tie into the electrical grid, called **grid-tied, utility-interactive, or grid-direct** (without batteries), essentially allows sharing. When more power than you are using is created, it goes out into the utility lines and to the nearest electrical load, probably to the neighboring home or office, displacing electricity from the fossil-fueled power plant. When more power is needed in your home or office than the panels are producing, extra juice is pulled from the power lines.

This ability to share a building's generated electricity has several benefits beyond bringing down costs and greatly improving the PV system's efficiency. For the average building owner who is probably not well-versed in the ins and outs of electricity, the biggest benefit is likely to be the fact that you never run out of electricity with a grid-tied solar electric system. Probably equally as important is that, compared with an off-grid system, a grid-direct PV system offers the near elimination of maintenance. It is no longer necessary to constantly check system voltage, worry about the kids using too much energy when you're away at work, or dodge the tangle of potentially lethal cables to refill the batteries with distilled water every month or two.

Utility-interactive systems' ability to share power also frees up PV system size. In off-grid systems, the amount of power-generating solar panels is tied into a formula based on the building's electrical requirements, average solar availability, and periods of cloudy weather. If the system is oversized, too much power is available to no benefit; if the system is undersized, residents are constantly agonized over turning off loads and cranking up a standby generator. By sharing power, PV systems can be any size, and considerations such as the size of the renewable energy budget and roof space can become the primary consideration. A building can put in a small system and displace just some of its power, or a large system can be installed that is capable of producing even *more* electrical power than the building consumes, an excellent goal for folks who strive to leave the world a better place than they found it.

The relative simplicity of grid-direct systems also means, at least in our opinion, that the electrical supply is more reliable than a stand-alone system. This might seem counterintuitive, since stand-alone systems can function while the grid is down and grid-direct systems without batteries cannot (more on this later). Nevertheless, stand-alone systems need maintenance; if the maintenance is not duly performed, they are subject to full or partial failure from a variety of sources—being drained down by long periods of cloudy weather or unnoticed loads, component failure from lightning strikes, and battery damage from neglect or excessive discharging. This lack of reliability can be ameliorated by the use of a fossil-fuel powered generator, but at additional cost and additional maintenance.

Last but not least, grid-direct PV systems are significantly less expensive and require far fewer components. This means that both the financial and energy returns are substantially better on installing a grid-tied PV system over an off-grid one. Financial return is improved because of the relative efficiency and simplicity of the grid-direct system, but also because the renewable electricity that is produced can sometimes be sold for a premium over the coal-based electricity bought from the utility (something we detailed in Chapter 3, "What's Appropriate for Your Budget"). Beyond financial return, different types of solar electric systems have an energy return on energy investment (EROEI)—often referred to as **net energy**—that refers to the energy profit that is reaped beyond what it takes to produce, install, and maintain them. Although net energy is correlated with financial return, it is a separate concept that limits itself to energy consumed versus energy produced over the lifetime of the system. Studies show the EROEI for solar electric systems is usually less than three years, meaning an amount of energy equivalent to the embodied energy in materials and manufacture is produced by the system before the fourth year. Because grid-direct systems use fewer components and operate whenever they receive sunlight—and especially because they function without energy-intensive batteries—their net energy can be much greater than off-grid systems whose heavy, lead-acid batteries typically need replacement after five to ten years. The positive net energy return of grid-direct systems goes a long way toward making solar electricity both renewable and sustainable in the long run.

None of this means you shouldn't have an off-grid solar electric system installed if that's appropriate for your home. It's also possible to have a grid-tied system that is backed up with batteries so it can function when the power grid is in a blackout or brownout situation. Such a hybrid system, usually referred to as **grid-tied battery backup**, can be a lifesaver for folks who need reliable access to electricity to perform necessary tasks—such as pumping water, running heating equipment, or keeping computer servers running—while gaining many of the advantages of grid-direct systems. Keep in mind that since off-grid and grid-tied battery backup systems are now in the minority of PV installations, you'll have to do a little more homework in finding a reputable and affordable contractor. This is especially true since these two types of systems require additional maintenance. Also, their additional complexity ensures that battery-based systems will always be more expensive to install and maintain than simple battery-free, grid-direct systems. For the remainder of this chapter, we'll examine how these different types of systems function and the responsibilities of the building owner in keeping the solar juice flowing so that you can make the best decision for your circumstances.

LIFTING THE WIZARD'S CURTAIN: SHEDDING SOME LIGHT ON HOW PV MODULES ARE MANUFACTURED

Beneath the veneer of magic are, of course, actual scientific and manufacturing processes that provide for the capture of sunlight and its conversion into electricity. While the functioning of a solar water heater, solar radiator, or solar oven are certainly impressive, most folks intuitively understand that these devices are simply well-designed insulated boxes that capture solar energy, turn it into heat, and move it to where it's needed. But the mysteries of solar electricity remain elusive.

How They Work

The photovoltaic effect (from the Greek *photo* for "light") depends on semiconductors, materials that are neither perfect conductors nor insulators, that allow for the regulated flow of electrical energy. They also depend on materials that easily allow for the transfer of electrons. In a PV cell, this is most commonly accomplished by "doping" a bottom absorber layer of silicon (the semiconductor) with boron (the positive layer). On top is a layer of phosphorus-doped silicon that acts

Figure 5.1. Solar electricity begins with the photovoltaic (PV) cell, which creates current and voltage when exposed to photons of light. PV modules are made up of many cells wired together. PHOTO BY ROGILBERT, WIKIPEDIA.

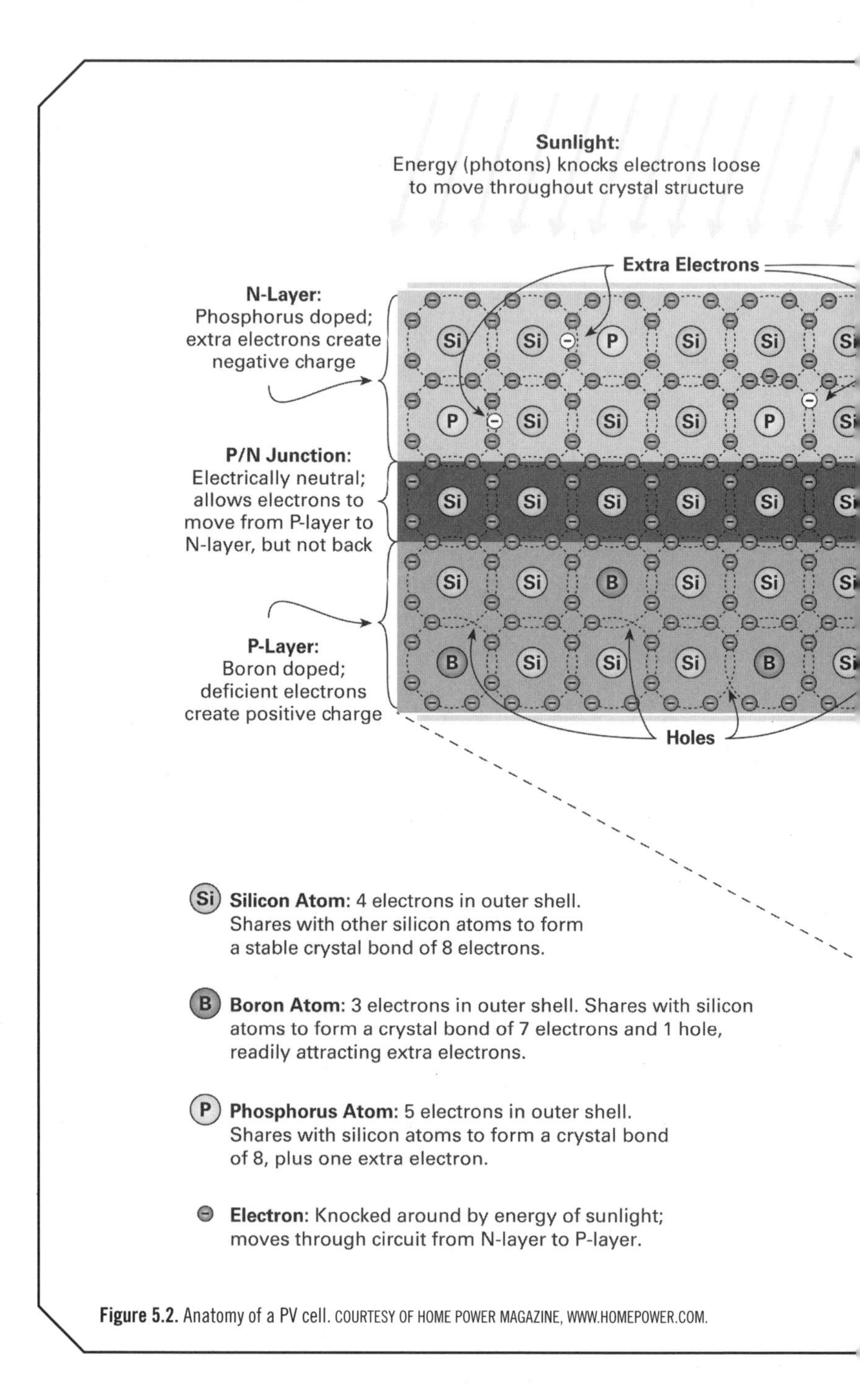

Figure 5.2. Anatomy of a PV cell. COURTESY OF HOME POWER MAGAZINE, WWW.HOMEPOWER.COM.

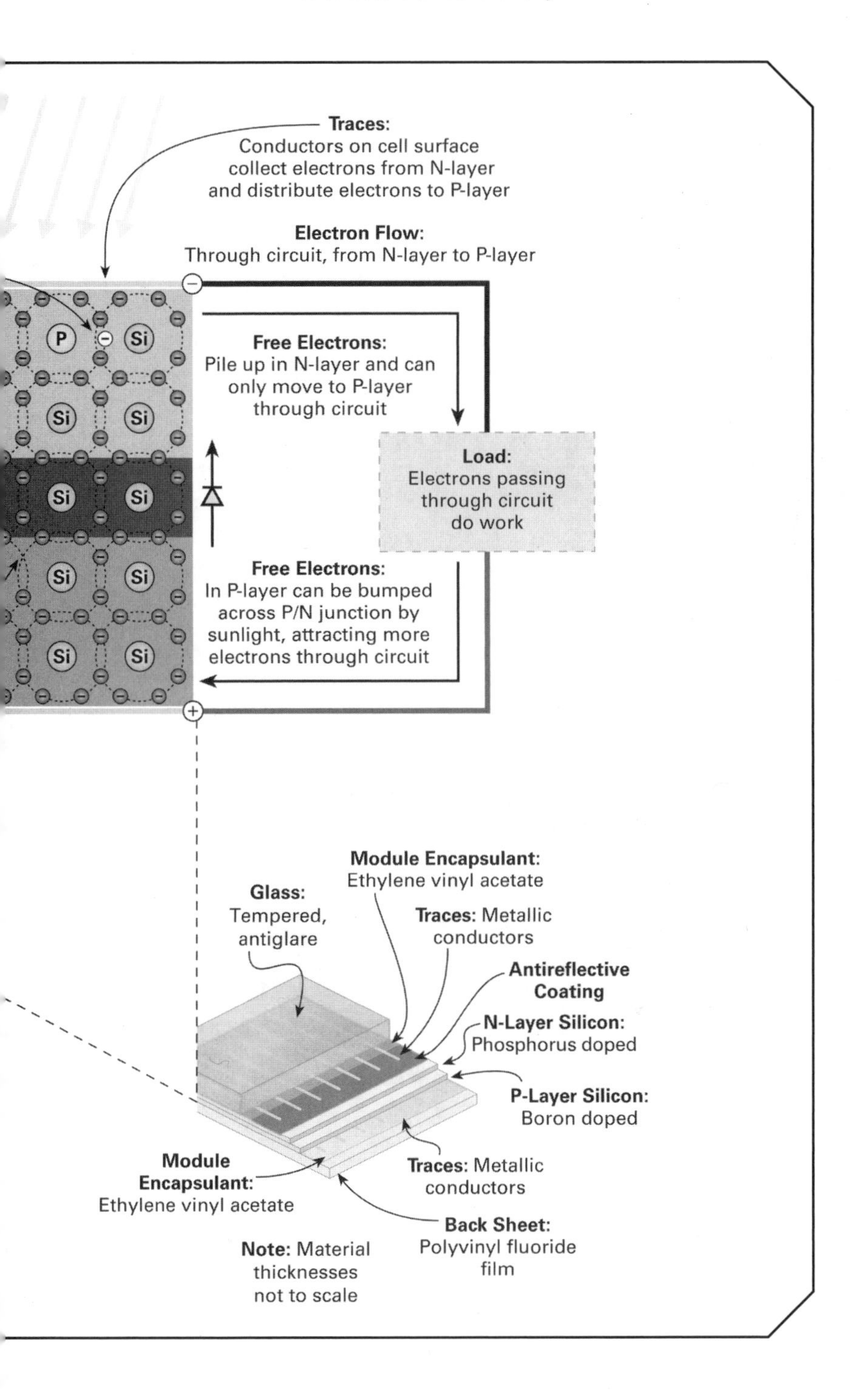
Traces:
Conductors on cell surface collect electrons from N-layer and distribute electrons to P-layer
Electron Flow:
Through circuit, from N-layer to P-layer
P
Si
Free Electrons:
Pile up in N-layer and can only move to P-layer through circuit
Load:
Electrons passing through circuit do work
Free Electrons:
In P-layer can be bumped across P/N junction by sunlight, attracting more electrons through circuit
Module Encapsulant:
Ethylene vinyl acetate
Glass:
Tempered, antiglare
Traces: Metallic conductors
Antireflective Coating
N-Layer Silicon:
Phosphorus doped
P-Layer Silicon:
Boron doped
Module Encapsulant:
Ethylene vinyl acetate
Traces: Metallic conductors
Back Sheet:
Polyvinyl fluoride film
Note: Material thicknesses not to scale

as the emitter (the negative layer). Photons of sunlight enter the PV cell and their energy is transferred to electrons, bumping them out of the electron shell of an atom and off into the electrical circuit. There is a static region in between the two layers known as the P/N junction that acts like a one-way valve (a diode) so electrons can only flow in one direction through the circuit. Lines of a highly conductive metal such as silver (often referred to as traces) are laid onto the individual cells and gather the electrons into the circuit. The potential difference in voltage—the pull of the electrons toward the positive layer—keeps the flow going, as long as sunlight is available to provide the energy to continue to knock new electrons loose. It is the energy from these freed electrons (not the electrons themselves) in the form of an electromagnetic field (EMF) that then flows through the circuits of the PV array and out into the wires as electricity.

How They're Manufactured

Most PV cells fall within two groups: crystalline silicon or thin-film. Crystalline modules are the classic variety, still make up over 80

Figure 5.3. Roof-mounted PV systems. The sleek-looking panels are unobtrusive once installed, and often impress friends and neighbors—as well as increase property values. COURTESY OF HONEY ELECTRIC SOLAR, INC.

percent of the market as of 2009, and are the most efficient (up to 20 percent). To make crystalline modules, you start with individual cells. There are two basic types of crystalline cells—monocrystalline and polycrystalline—which have slightly different manufacturing techniques and thus different electrical characteristics. For monocrystalline cells, a seed crystal is pulled through highly refined silicon melt doped with boron at around 2,500°F, and a boule (cylindrical crystal) about five to six inches in diameter is drawn out; this is the positive layer. After being cut into razor-thin circular wafers, it is coated with phosphorus and then heated to allow the phosphorus to seep into the silicon and adhere; this is the negative layer.

Polycrystalline cells cast molten silicon doped with boron into a large square block, and then cut this into razor-thin wafers. The process is simpler, but results in lower efficiency due to the randomly oriented shardlike crystals (which means polycrystalline modules look sparkly). The cell is then coated with the phosphorus layer in the same manner as monocrystalline cells.

The cell is shiny gray at this point and highly reflective, a serious problem since mirrors make poor solar absorbers. Several layers of antireflective coating reduce this reflectivity to less than 4 percent and change the appearance to bluish-black. To make modules, individual cells are squared off, overlaid, and connected with a conductive grid (typically silver) to carry electrical current. Modules are then encapsulated for weatherproofing, usually with tempered glass. Tedlar (a polyvinyl fluoride film) is frequently used for module backing, although glass is also used. A laminate (ethylene vinyl acetate) seals the front and the back of the cells to the glass and Tedlar. Modules are then enclosed in a mounting frame, usually aluminum.

The Other Types

Beyond the monocrystalline and polycrystalline types described above, there are also ribbon silicon cells and thin-film cells. Ribbon silicon cells are a polycrystalline type made using string-ribbon technology. Instead of casting, the positive layer of silicon is drawn out of the silicon between parallel strings. This produces a thinner cell that doesn't require sawing.

Thin-film cells can be made of deposited layers of doped silicon rather than grown crystals, or indeed they can use a different semiconductor entirely (such as cadmium telluride). This deposition can be applied by spraying, vapor application, or screenprinting, and can offer a significant cost savings over crystalline technologies. A transparent conductive oxide is applied and serves as the conductive path. Because thin-film cells lack fragile crystals, they can be applied to flexible materials—everything from laptop bags to roofing shingles. With many new startups and lots of research and development, expect to see more thin-film modules on the market at bargain prices per watt. They do, however, suffer from substantially lower efficiencies (4 to 14 percent). Because of the lower efficiencies, more space and racking is needed for thin-film systems.

How the Magic Happens: The Nuts and Bolts of Residential Solar Electric Systems

Before we explain the basics of your future PV system, we must emphasize one key element that will make your system much less of a financial burden. As the most expensive type of solar technology, the return on investment in solar electricity benefits the most from energy conservation and efficiency. Simple measures such as eliminating phantom loads and avoiding turning electricity into heat (for example, a clothes dryer) mean that you can offset a much larger proportion of your electric bill with a smaller PV system. These topics are beyond the scope of this solar guide (we explore them in detail in our previous book, *The Carbon-Free Home*), but conservation and efficiency are inextricably linked and are ignored at the homeowner's expense. For more detail on analyzing your current electrical consumption, see "Understanding Your Current Energy Usage" in Chapter 2.

Although there are differences in how the three types of solar electric systems (grid-direct, grid-tied battery-backup, and off-grid) work, the fundamentals are the same. We'll begin with a general overview of the components of a residential PV system, and then break out what's different about each type.

The first and most important component is the **PV module** itself. Each PV module, colloquially referred to as a PV panel, is composed of many smaller PV cells, each of which is made of a semiconductor like silicon (with boron and phosphorus added), that generates an electric charge when exposed to light (see the previous sidebar "Lifting the Wizard's Curtain"). PV modules are rated (in a test lab) by the amount of power, measured in watts, that they produce when exposed to a standard set of sun and temperature conditions. A typical residential-sized module will generate 150–250 watts, and be somewhere in the neighborhood of three by five feet and contained in an aluminum frame about an inch thick. PV modules are added together to create a **PV array**. The power output of the PV array is its DC wattage, which is simply the sum of the output of each of the modules in the array. Since there are other fixed costs associated with installing a PV system (the PV array plus other components, listed below), cost per installed watt tends to decline with more modules, which is why you rarely see just one PV module up on someone's roof. A common residential system size is 3,000 direct current (DC) watts—also referred to as 3 kW—although sizes will vary greatly. A 3 kW PV system will produce approximately 3 kWh of DC electricity per one hour of clear midday sun. The actual AC production number will always be somewhat lower than these numbers due to inefficiencies, line losses, and other environmental factors.

PV panels are not extremely heavy (thirty or forty pounds), and since they cover a large area, you might guess that they make pretty good sails. That's why the next thing you'll need is some way to hold your PV array in place. This can either be done with racking that is attached to your roof or, if there's a more appropriate place in your yard, mounted on poles or piers sunk into the ground. Racking is a substantial cost, and minimizing racking while ensuring a secure installation not only brings down costs but often improves the final look. New racking systems are

Figure 5.4. Ground-mounts make sense for larger systems that won't fit on the roof, or in places where the yard has the best solar window. COURTESY OF HONEY ELECTRIC SOLAR, INC.

Figure 5.5. Pole-mounts are similar to ground-mounts, but can be seasonally adjusted more easily, and are often seen on ranches or other wide-open land. COURTESY OF HONEY ELECTRIC SOLAR, INC.

constantly becoming available, so ask your installer if they are up to date on the many choices, especially if your roof is covered with less common materials like metal or tile. Generally speaking, a roof mount will be a more efficient (meaning cheaper) use of material and labor than a pole- or ground-mount, assuming you have an unshaded, non-north facing roof.

Many people are curious about tracking mounts, which are ground-mounted arrays that move to follow the sun's path through the sky. Keeping panels precisely perpendicular to the sun's rays allows for the greatest power production, so trackers can add a significant percentage of energy production to certain systems. The question to ask is, will the extra energy production offset the added cost and maintenance of a tracker? Frequently the answer is no, and a better investment is more PV modules! Your solar installer will be able to help you do a cost/benefit analysis.

The electricity coming out of PV panels is known as direct current (DC) electricity. This is in contrast to the type of electricity that comes over the grid in sine waves, called alternating current (AC), that powers everything in your home. Back when Nikola Tesla and Thomas Edison were battling it out at the dawn of the electric age, DC current (backed by Edison) was a candidate for being the primary method of transmitting electricity over power lines. But Tesla's favored alternating current electricity was more easily converted between high and low voltages, making transmission less expensive. This fact, along with Tesla's strong financial backing from Westinghouse, resulted in alternating current winning the so-called war of the currents. What this means for solar enthusiasts is that all power generated by PV panels must be changed from DC to AC current with an **inverter** to make it usable by standard household AC appliances and the grid. There are also various disconnect switches and overcurrent protection mechanisms (fuses and breakers) in the system to ensure safety.

Beyond these basic components, every home has an **AC main service panel**. From this panel, the electrical loads in the house are served through circuits made of copper-wire conductors, each of which has its own breaker in the panel. The most typical PV

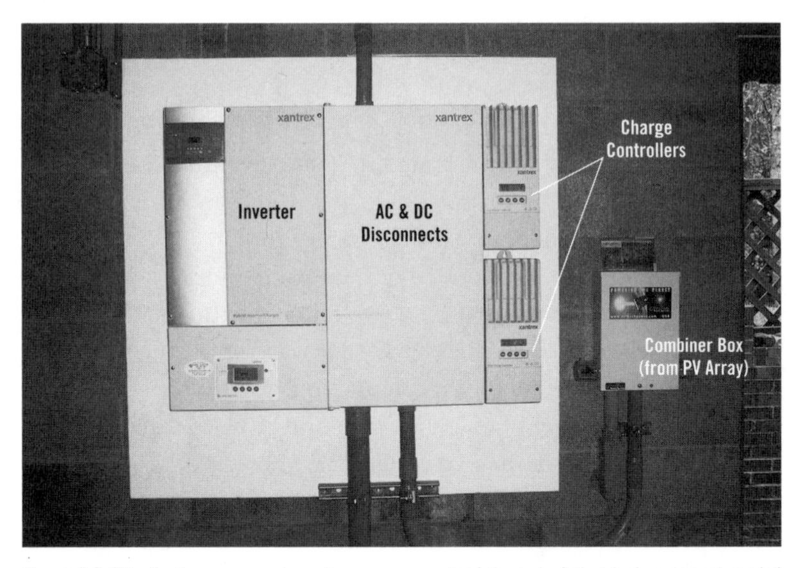

Figure 5.6. This Xantrex power system shows one example of the sort of electrical equipment needed with a battery-backup PV system, including charge controllers on the far right, a disconnect (breaker) box in the middle, and inverter on the left. COURTESY OF HONEY ELECTRIC SOLAR, INC.

installation will have the inverter wired to what appears to be identical to any other breaker in the panel. But instead of taking electricity from the panel, this circuit feeds electricity into the panel to be sent out to the loads. In the case of a solar home, an AC service panel may interact either with the grid (if the system is utility interactive) or just serve the loads of the home (in an off-grid scenario).

The three types of solar electric systems (grid-direct, grid-tied with battery backup, and off-grid) vary substantially in detail and cost. Before we begin to look more closely at how each type of system works, it's worthwhile to understand these significant differences. Table 5.1 provides a rough breakdown of costs for similarly sized systems, along with the major pros and cons and system components. For more detail on all the financial considerations of the different types of solar, check out Chapter 3, "What's Appropriate for Your Budget."

Figure 5.7. The Fronius grid-direct inverter (on the left) connects the house's main electric panel (on the right) to the PV array on the roof. There is a breaker in the panel labeled "solar," which is the interconnection point to the utility grid. COURTESY OF HONEY ELECTRIC SOLAR, INC.

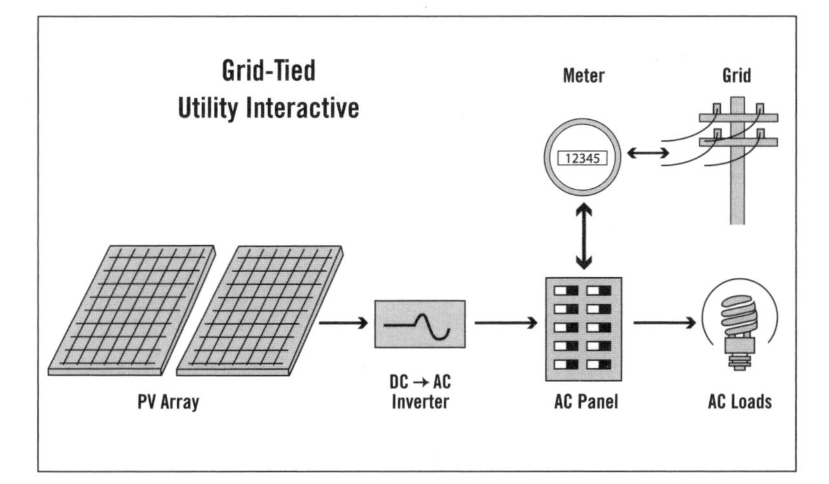

Figure 5.8. Diagram of a grid-direct (utility-interactive) PV system. The PV array produces DC electricity, and the inverter converts DC to conventional AC electricity. Whatever electricity is left after satisfying the loads in the home or office is sent out to the grid and flows to the nearest neighbor's electrical load. Grid-direct systems are designed around average consumption habits, not for specific loads.

Table 5.1. Average Costs for Different Systems (2 kW and 4 kW; grid-direct, grid-tied battery backup, and off-grid) Prior to Incentives.

PV Array Size	2 kW	4 kW	Pros and Cons
Grid-Direct	$12,000 to $22,000	$20,000 to $40,000	**Pros:** Simple, efficient, less expensive **Cons:** No power during grid failure/ blackout/brownout
Grid-Tied Battery Backup	$20,000 to $36,000	$32,000 to $56,000	**Pros:** Power available for specified loads during power outages **Cons:** More maintenance, cost, components, and space needed
Off-Grid	$20,000 and up	$32,000 and up	**Pros:** Functions in remote locations where access to grid might be too costly **Cons:** Similar to battery backup, but also needs backup power source during cloudy periods

Details of the Three Types of Solar Electric Systems

Beyond the basics outlined in the last section, each system has distinct components that allow it to function in its particular niche. The vast majority of new solar electric systems (over 90 percent) don't rely on batteries, and these are the first systems we'll discuss. But the latter two systems might be appropriate for your home if special circumstances apply and you're willing to shell out the additional dough.

Grid-Direct Solar Electricity

Tying a solar electric system into a home that is attached to the power grid is the easiest and cheapest method of installation. Simple and efficient, whether retrofit or new construction, a grid-direct PV system will displace some, all, or more electricity than your home currently consumes. Figure 5.8 shows a diagram of how the basic components of a grid-direct system relate to each other.

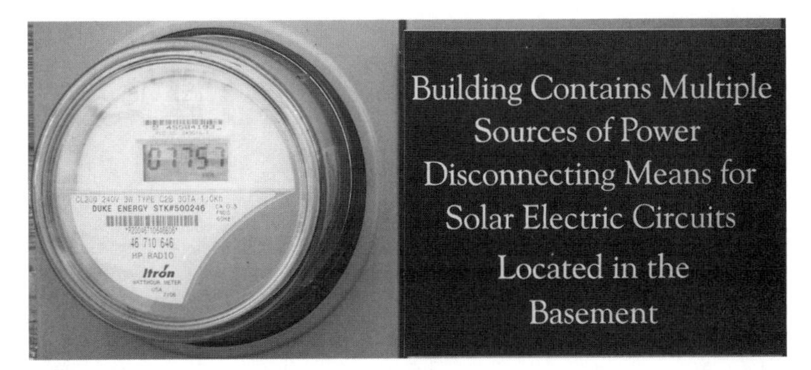

Figure 5.9. In some states, second meters are installed to measure electrical flows out to the grid. Other locations measure both incoming electricity and outgoing electricity in the same meter.

Here's the nitty-gritty of how it works: A residential PV array generates DC electricity that is inverted to standard AC electricity at the household inverter. There are myriad sizes of grid-direct inverters on the market (inverters are sized in watts, the same as PV modules). You might need just one central inverter for your array, or perhaps multiple inverters if you are installing a large array. Another possibility is a micro-inverter, which is sized precisely for the generating capacity of one PV module. Micro-inverters can be strung together in an AC branch circuit, and allow for PV modules to be added to the array one or more at a time. Another benefit of micro-inverters is that instead of a small bit of shade taking out a large proportion of the array's potential (as can happen with larger central inverters because of the way modules are electrically connected in series strings), shaded modules instead shut down production individually, allowing the remaining sunny modules to continue to kick out the juice.

The generated AC electricity from the inverter flows to an AC service panel with breakers, and from there to AC circuits within the home that require power. If more electricity is generated by the PV array than is consumed by the home, the additional electricity flows from the main service panel to the utility meter, where it spins the meter backward (digital or analog meters can both be made to work this way), and then out to the electrical grid. When there's

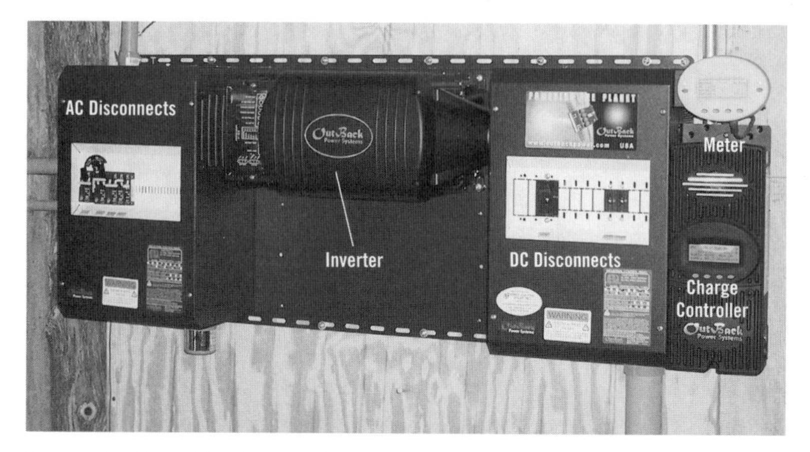

Figure 5.10. Off-grid PV systems are designed to provide energy for specific loads, whether a well pump or a whole cabin. The voltage of the battery bank must be regulated by a charge controller, which prevents overcharging of the batteries and can provide other functions as well—such as data-logging, voltage conversion, or preventing the batteries from draining too low. COURTESY OF HONEY ELECTRIC SOLAR, INC.

not enough power being generated by the PV array to cover the residential electrical loads, power is taken from the electrical grid. The meter registers this positive flow of electricity from the grid. This classic configuration is known as **net metering.** By spinning the meter forward and backward, you are only charged for net electrical consumption, which can be zero! If you actually make more energy than you consume over a period of time (usually a year) rules vary on whether you will be compensated or if the energy is "donated" to the utility.

Net metering (when a meter spins backward and forward, measuring only *net* production or consumption) is the most common way to tie in grid-direct systems, but variations abound on how to add the capability of measuring input (consumption) and output (production) separately, based on state laws and utility regulations. A relatively common alternative is to have two meters, one for total production and one for total consumption, instead of having one meter spin forward and backward. Don't fret over these details, but instead invest time finding a trustworthy installer (see Chapter 4, "Getting Ready for the Installation") who knows the best setup for your area.

Beyond these differences, grid-direct PV is a straightforward affair, as the name implies. If the utility grid goes down, as in a blackout or brownout, the PV system inverter also shuts down—and no power can flow from the array to the house. Without batteries in a system, no power can be used during a blackout, as the variability of incoming sunlight makes the power source extremely unstable (imagine your TV turning off every time a cloud hovers overhead). In addition, utility regulations mandate that grid-direct inverters can't send power to the grid during a blackout to ensure the safety of linemen (this is called **anti-islanding**, meaning the inverter won't create an island of power).

The only homeowner maintenance relates to making sure the inverter is humming along without errors, and that the PV array stays relatively clean, which might mean hosing off pollen or dust (pressure washing is not recommended). Regions with regular rainfall don't need to worry much about cleaning off the array, but dusty regions might need regularly scheduled cleanings—and there are companies popping up that specialize in cleaning arrays. Your solar installer should be able to provide a routine maintenance contract to do a system check-up once every year or so to make sure wires are screwed down tight and there is no visible damage to any wires, modules, or conduit.

Grid-Tied Battery-Backup Solar Electricity

PV systems backed up with batteries offer redundancy for people who have electrical loads that they want to function even when the grid offers no service. Of course, the perception of this need will also depend on the reliability of the power grid in your region. There are areas of the country that see few debilitating storms, where the stability of the grid is pretty much taken for granted, and where there might be little reason to invest in battery backup that would be rarely used except in exceptional circumstances. However, more isolated locations that are subject to harsh storms or unstable grid voltages that result in relatively frequent power outages might benefit greatly from having their PV system support important loads during such an event. This is especially true if loads do things like maintain access to drinking water or enable the primary heat source.

Once batteries are involved, the electrical situation becomes markedly more complex. Batteries deal with direct current just like PV modules. Batteries are subject to damage if overcharged, and need voltage regulation. This is done with a **charge controller**, also sometimes referred to as a battery regulator, that is positioned between the PV array and the **battery bank**. The battery bank is one or more batteries that store an amount of energy to service the **backed-up loads subpanel** in the event of a utility outage (energy flows via the inverter). Circuits that service the backed-up electrical loads are fed from this subpanel (which is automatically isolated in the event of a grid power outage) and pull electricity from the inverter, usually from the PV array or grid but via the battery bank during outages (at standard household AC current). For power outages, the battery bank is generally sized to power the chosen loads for a day or two while the grid is down, although the length of time energy is available depends on the loads, the size of the battery bank, and the resident's power usage. The inverter makes sure that the backed-up loads remain electrically isolated during a grid outage; once again the inverter is "anti-islanding," meaning the inverter won't try to feed AC power back into the grid while the utility lineman may be working to recover power.

Beyond this subpanel branched off from the inverter, the battery-backup system functions the same way as a grid-direct system does. During power generation (i.e., when the sun is shining), the AC power goes to loads in the house. If there is still extra juice, it is then measured by the utility meter and proceeds out to the grid. When the residence's electrical draw is above what the PV array is producing, power is pulled from the grid (and feeds all loads, including the backed-up loads). The batteries aren't normally cycled, but wait on standby at full capacity while anticipating a grid disruption. This prolongs their life and ensures the maximum availability of energy in the event of a disruption.

Off-Grid Solar Electricity

Back when solar electricity was still a novelty used almost as often on satellites as it was in residences, all systems were stand-alone (i.e., off-grid). Generally speaking, both the homeowner (who valued their sustainability and energy independence) and the

Figure 5.11. Batteries used in off-grid and battery-backup systems need to be stored properly and safely. They are potentially dangerous on three levels: accidentally shorting during maintenance, leakage or spray of sulfuric acid, and hydrogen offgassing. COURTESY OF HONEY ELECTRIC SOLAR, INC.

BATTERIES: BOON OR BOONDOGGLE?

PV systems with batteries introduce not only additional components at additional cost of purchase and installation, but also increase the maintenance level compared with a grid-direct system. Batteries are generally lead-acid, with no substitute such as lithium-ion batteries or a household fuel cell likely in the medium term. A lead-acid battery bank takes up substantial space and can weigh hundreds or even thousands of pounds. Moreover, it represents a triple safety hazard. Battery wires and terminals are potentially extremely harmful if short circuited—meaning if the positive and negative are directly connected by something that conducts electricity. The sulfuric acid contained in the batteries is caustic and can produce third degree chemical burns. And finally, the hydrogen produced as batteries charge is a potential explosion hazard. For these reasons, the battery bank must be secured in a battery box in a safe, out-of-the-way location—and be well-vented. Sealed lead-acid batteries can solve the issues of sulfuric-acid dangers and hydrogen production, but at additional cost and reduced life span. They also eliminate the need to add distilled water, and so greatly reduce maintenance. Nearly every part of a battery can be recycled, though!

utility (who wanted nothing to do with uncontrolled distributed power generation, especially from cobbled-together, homeowner-installed PV systems) were happy with this arrangement. As we discussed in the introduction, times have changed—and off-grid solar electric systems are now the odd man out. But for remote locations, or for those who simply don't trust the longevity of the grid in the face of dwindling fossil fuel supplies and exploding national debt, off-grid PV systems may still make sense. Also, for devices that use DC and can be powered directly from a PV module or array (called **PV-direct**), such as a pump or fan, keeping the system off-grid greatly reduces both cost and complexity.

Off-grid systems are the most variable of solar electric systems, and encompass everything from farmers using PV modules to directly pump water into their fields to much larger systems capable of powering residences or even small communities, offices, and hospitals. For the homeowner, a standard setup is shown in Figure 5.12. Just like the battery-backup grid-tied system, off-grid systems charge batteries with the DC current produced by the PV array, regulated by the charge controller (see Figure 5.10). Power, either from the PV array or batteries, then goes to the inverter where it is switched to alternating current.

For most residential or larger off-grid systems, a secondary source of power is necessary to ensure sufficient energy supplies during times of low solar output. Typically this is supplied by a gasoline, diesel, propane, or natural-gas powered generator that can be tied into the inverter. In this circumstance, the inverter can also act as a battery charger, and is more technically called an **inverter-charger**. This charging role is distinct from the charge controller that regulates the interaction between the PV array and the batteries.

Some off-grid enthusiasts scoff at the idea of a fossil-fueled backup power source, believing you should live strictly off your solar budget.

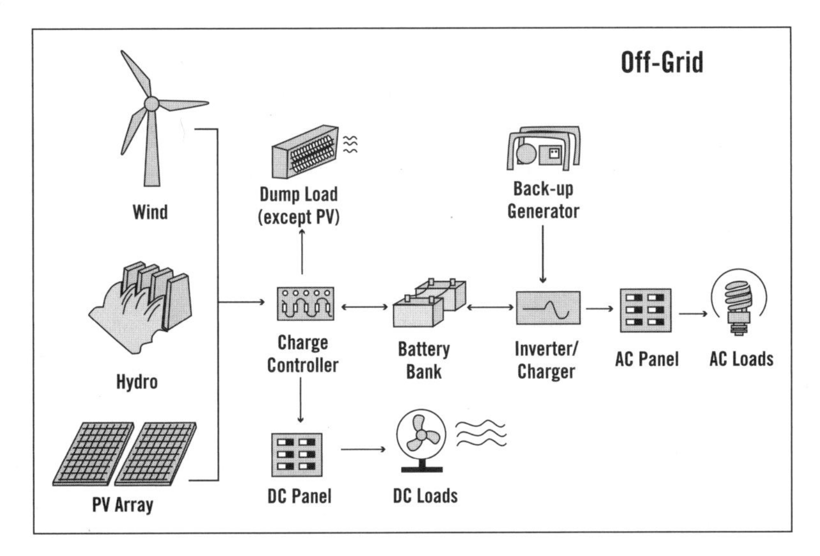

Figure 5.12. The level of complexity for off-grid PV systems is much higher than for grid-direct, and the results are higher costs and maintenance. Unlike grid-direct PV systems, off-grid systems are designed to carefully balance isolated supply with very specific loads, whether a whole house or a well pump, making system sizing difficult and creating the need for electricity storage and regulation.

But while a generator is burning dirty energy, and making some noise pollution to boot, tying one into an off-grid system can have the benefit of reducing the size of the battery bank, because you don't need to store as much energy for stretches of low solar output. Batteries, of course, are made using fossil fuels—and lots of them. So if your off-grid PV system can halve the size of the battery bank, that's not too shabby. For more on the trials and tribulations of batteries, see the "Batteries: Boon or Boodoggle?" sidebar on page 82.

Off-grid systems are often located far from the grid, and in such rural locations other renewable energy sources might also be available. Both grid-direct and off-grid systems offer the opportunity of tying in a wind-based or hydro-based source of renewable energy to supplement the solar source. Oftentimes, and especially for off-grid systems, two types of renewable energy work synergistically together. For example, wind energy is often available in the early morning and late evening, when the solar resource has set. A small hydro system (referred to as micro-hydro) might provide twenty-four-hour energy supplies—with maximum output of electricity in the winter when flow rates are generally higher, and less output in the summer when the solar resource is coming on strong.

To complete the workings of the typical off-grid system, after leaving the inverter the generated electricity continues on to an AC panel with breakers. From there, all the circuits of the residence are served.

Conclusion

Solar electricity is amazing stuff, nearly universal in availability and applicability. Depending on your location and needs, you can install a simple grid-direct system, a grid-tied system with battery backup in case of frequent brownouts or blackouts, or a stand alone off-grid system in remote locations. Systems are almost infinitely scale-able, and can provide any percentage of your electricity, with your solar window and budget being the constraints!

Although production may vary during the year, large amounts of solar energy are available throughout North America for all types of solar installations. Before committing to a solar electric system, however, be sure to read the next chapter on solar hot water—usually the easiest and most cost-effective method of utilizing solar power.

CHAPTER 6

Solar Hot Water

If you've ever tried to wash dishes or shower in freezing cold water, you'll agree that hot water is a necessity year-round. Conveniently, the sun is an excellent heat source, shining year-round. Since heating water takes lots of energy, currently supplied primarily by fossil fuels that are both polluting the Earth and finite in supply, it makes sense to think about supplying our hot water needs with solar energy.

If the technology existed to convert our automobile fleet to run almost entirely on renewable energy with minimal cost, most Americans would be overjoyed. The equivalent technology already exists *right now* for a similarly nefarious polluter, our domestic water heaters—typically the second largest energy user in the home and office after heating and cooling. Solar water heaters can be installed for just a few thousand dollars after tax credits, and provide on average 70 to 80 percent of the needed energy to heat our water. Solar and hot water go together like salt and pepper, peanut butter and jelly, and Tweedledee and Tweedledum. Every home in the world should have a solar water heater on it. If you only have the space or budget to put one type of solar technology on your home, it should almost certainly be a solar water heater.

Exceptional Advantages of Solar Water Heaters

We've told you about what wonderful technology solar water heaters are, but let's break it down in more detail so you can understand why an investment in solar water heating should most likely trump not just other solar installations, but also investments in many other types of financial vehicles such as savings accounts, stocks, and bonds.

Q. Will my water be hot enough with a solar water heater?

A. Absolutely. Solar water heaters have backup power sources to make sure your water stays exactly the temperature you want it to during cloudy weather. During sunny weather the water can get so hot that systems come with an automatic mixing valve to cool it down!

Q. Can't I just use solar electricity to run my electric water heater? Then I'll only need one type of panel!

A. Yes, but it would be at least three times less efficient and much more expensive than using each type of solar device for its designed purpose.

Q. How much of my hot water will a solar system make?

A. It depends on the size of the system, the time of year, your climate, and how much hot water you use. But generally speaking, between 60 and 90 percent of your hot water will come from a solar water heater.

Q. I have a fairly new water heater—would I have to replace it if I get a solar water heater installed?

A. No, the system can be configured to prefeed your existing tank with solar-heated water, so there is no need to worry about prematurely replacing a perfectly good tank.

Q. Solar water heaters seem great, but I'm the only resident in my home—and I travel a lot. Does it really make sense for me to spend the money on one?

A. Maybe, maybe not. Definitely read through this chapter and research your local incentives (including utility-sponsored ones). It may make sense to consider an on-demand (aka instantaneous) water heater that only turns on when hot water is needed and eliminates the standby losses associated with hot water tanks. Then you can dedicate your solar budget to solar electricity instead, which can produce usable renewable energy (because it can be shared with the grid) whether you're home or not.

Turning solar energy into heat is a much simpler process than turning it into electricity, and for that reason alone heating things up with the sun will likely always be more efficient than generating electricity. At present, overall efficiencies for solar water heaters are around 60 percent, while photovoltaic systems average around 15 percent. What that efficiency number means is what percentage of the power available in sunlight is captured by the technology and converted to useable energy. To capture the same amount of energy from PV modules as from solar thermal collectors *requires four times as many panels!* And unlike solar heating options (discussed in Chapter 7), solar hot water is used year-round. So even though both use solar thermal technologies, solar hot water captures about twice as much energy annually as a typical solar heating installation.

The cost of installing solar hot water usually varies between about $3,000 and $9,000, depending on the type and local labor costs. Federal tax credits good through 2016 knock 30 percent off this number, and state tax and utility incentives can reduce the cost substantially more (as discussed in Chapter 3, "What's Appropriate for Your Budget"). For a typical household spending $300 to $600 annually on hot water, solar hot water will usually pay for itself in five to ten years—and this is for a system with an expected life of around twenty-five years. Such a payback often translates into an annual return on investment of 5 to 10 percent or so, and since this is a savings rather than actual income, it means the investment return is *before* taxes—unlike other investment income like stocks and bonds. This is a respectable rate of return given solar hot water's long track record and decades of dependability.

Solar water heating has a few additional advantages that usually make it the premier choice for bringing some sunshine into your home. The lower cost compared to PV makes for a lower bar to entry, and its exceptional positive investment return make it an easy sell to the local banker for those who need to borrow money to make their solar investment happen. And for folks with an imperfect solar resource, solar hot water's efficiency is exactly proportional to the amount of sunshine the collectors receive, unlike PV's disproportionately sharp drop in production from even small amounts of shade.

Types of Systems and Panels

A variety of effective solar hot water systems have been developed over the last few decades and centuries, and the myriad choices can be confusing and overwhelming. We'll sort out which options work best for your situation. Be forewarned that the various component and system setups are numerous, and listening to an installer talk about all these choices without any background understanding can make your head spin. We'll break it down for you as simply as possible, but keep this book handy in case you need to refresh your memory!

Also keep in mind that each contractor is likely to have their own preference for which type of system they want to install, whether because they know it works best in your climate or they get a deal on a certain type of water heater or collector from the manufacturer. If you do a thorough job picking a contractor (see Chapter 4, "Getting Ready for the Installation") then you should feel comfortable following their advice.

Often it will be the case that more than one type of system or collector is equally suitable, but gaining a basic understanding of the way each system works will allow you to ask your contractor intelligent questions—and of course explain how the thing up on your roof works to interested friends and neighbors. Regardless of the type of solar water-heating system, a backup heat source ensures that hot water is available even when the sun is not.

The Big System Difference: Direct Gain or Heat Exchange?

Heating water with the sun is a simple affair on the face of it. A garden hose left in the sun on a summer day will produce water so hot as to burn the skin. But scaling up to heat most of the water for a typical household (and without burning anyone) is a bit more complicated. The biggest factor in determining how complicated your solar water heater needs to be is how cold it gets where you live. Water (of course) freezes, and when it does it expands up to 9 percent with a force that can exceed 100,000 psi. Even very stout metal piping such as steel or copper will eventually split when such force is exerted upon it.

Where nighttime temperatures rarely fall below freezing during winter, a household's hot water can be heated via **direct gain** from the sun. Direct gain simply implies that the actual hot water that will be used in showers, laundry, and kitchen sinks is brought up to a sunny spot (usually on the roof), heated up, and stored in a tank for later use. The tank can be up on your roof in an insulated and glazed black box where the tank itself is heated. Or the water can be heated by one of the two other types of solar thermal collectors (evacuated tube or flat-plate, discussed in detail below) and stored in a tank somewhere in the home or basement. Both of these systems maintain simplicity (and hence keep costs down) by using household water pressure to move water.

A **batch collector** with the hot water actually stored on the roof (also called **integrated collector storage**) is the simpler of the two direct-gain systems since it requires no additional solar collectors, and thus can be installed easily and inexpensively. The major disadvantages of storing the heated water in a tank on the roof are both the additional weight applied to a relatively small area of roof and the fact that the tank loses heat when the sun is not shining. Typically batch systems perform best when the majority of hot water use is later in the day, while direct gain systems with an indoor tank and roof-mounted collectors perform better for hot water use early in the day, since the tank can be better insulated. A common threshold for installing either type of direct gain solar water heater is for homes in USDA hardiness zone 8 or higher, although local microclimates may prohibit it.

Where freezing weather is a concern, damage to pipes, collectors, and other exposed components is prevented by using a system of **heat exchange**. Heat exchange can function in one of two ways. The first method is a **closed-loop** (or **pressurized**) **heat exchange**, whereby a nontoxic food-grade propylene glycol that has a much lower freezing point (down to -50°F) is heated in solar

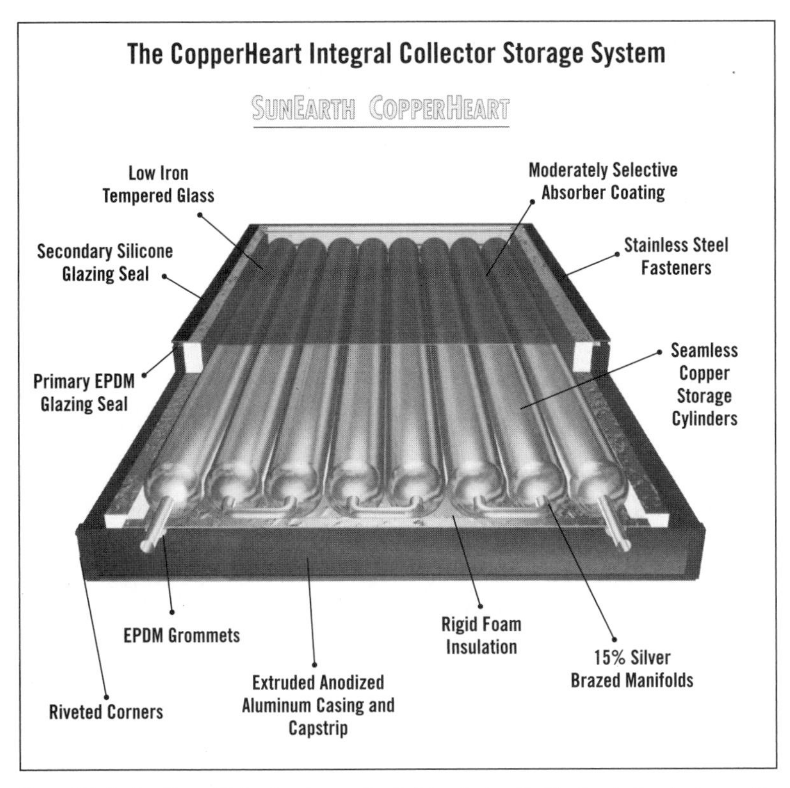

Figure 6.1. For locations with few freezing worries, batch collectors that heat water in a tank up on the roof make simple and inexpensive solar water heating installations. COURTESY OF SUNEARTH, INC.

thermal collectors and then pumped under low pressure to the hot water tank. This heated liquid then transfers its heat to the water in the tank by moving continuously through a heat exchanger (which can be coils around the bottom of the tank or another design like an external heat exchanger) and then back up to the collectors to be reheated. The closed-loop name refers to the fact that the fluid is these systems is closed off and separate from household water systems (so alternately, systems that circulate household water are sometimes called **open-loop**).

The second method for freezing climates is a **drainback heat exchange**. This involves pumping a liquid, often simply distilled water or water mixed with a small amount of propylene glycol, into the collectors when the temperature at the panels exceeds the temperature in the water heater. The unpressurized fluid circulates through the panels and heats the household water via a heat-exchange mechanism. The main difference with a drainback system is that when the temperature of the tank exceeds that of the collectors, the system stops pumping and the fluid "drains back" into a small additional reservoir tank in a heated location that prevents freezing of the liquid. When the sun shines again the next day, the fluid is pulled from the tank and the process starts all over.

Choosing the Collector: Flat-Plate or Evacuated Tube?

Once you've got an idea of whether you need a direct-gain or heat-exchange setup, the next big variable is the type of solar thermal collector that will heat your liquid. Keep in mind that a batch solar water heater warms the water in the tank directly on the roof, but for separate tank systems (whether they are direct gain or heat exchange) solar collectors are required.

Flat-plate collectors are basically insulated, glazed black boxes with small-diameter copper pipes running through them. A typical size is four by eight (or ten) feet, and one to three panels are commonly used for the average home installation (for domestic water use, not for space heating). They have been around for over a century and represent a time-honored and effective method of solar heating a fluid, and are the most common type of solar ther-

Figure 6.2. The most common solar water heating installation in colder climates utilizes flat-plate collectors with copper tubing to capture solar energy. COURTESY OF SUNEARTH, INC.

Figure 6.3. Evacuated tubes are an alternative to flat-plate solar hot water collectors; they may perform better in some climates. This installation is actually hidden from street view, angled to face south on a low-slope north roof. COURTESY OF HONEY ELECTRIC SOLAR, INC.

mal collector. **Evacuated tube collectors** (**ETs**) are a newer technology developed in the late 1970s that relies on a sealed glass tube, typically four to six inches wide by seven feet tall, that is evacuated of air for a very high insulation value. An average residential system will have in the range of twelve to twenty-four tubes. Individual tubes are linked up in series, connected to an insulated copper header that the heated fluid runs through and then back to the water heater.

In most cases, either flat-plate collectors or ETs will function well and provide a majority of a home's hot water needs. All else being equal, flat-plate collectors have historically been less expensive and have a great track record. They are likely the most economical choice, although their heavy copper component makes them susceptible to price fluctuations in this material. Beyond cost, there are a few things to consider that might help with your decision—assuming you can find an installer who is familiar enough with ETs so that you actually have a choice. You might go with ETs under one or more of the following situations:

- Very Cold Climate: ETs are substantially better insulated than flat-plate collectors, and perform better during sunny cold weather (generally, where daytime temperatures do not rise above freezing for much of the winter).
- Less Than Ideal Siting: ETs can be rotated from side to side, making for better collection if the roof angle is more than 30 degrees off of true south, or if the morning and afternoon solar window is shaded. ETs also perform more evenly over a greater range of sun angles and rack orientations.
- Aesthetics: ET and flat-plate collectors have substantially different looks, and one style may be more visually appealing to you. Look at Figures 6.2 and 6.3 for comparison. Also, flat-plate collectors are unsealed, which can mean eventual buildup of condensation on the glass. While this doesn't greatly affect performance, it doesn't look as sleek as ETs (they don't suffer this problem).
- Snow and Wind Loads: ETs that are installed parallel to the roof are a victim of their superior insulation when it comes to snow. Since they lose less heat, they have a harder time melting off snow accumulation. However, when installed on a rack and elevated above the roof, ETs' round shape often allows snow to pass through the gaps and reduces snow accumulations compared to flat-plate collectors. Also, racked ETs have lower wind loading than racked flat-plate collectors—and weigh less, too.

A More Detailed Look at How the Different Systems Operate

To satiate your curiosity, and as a reference for both comparing different types of systems before hiring an installer as well as understanding how your system works once it's up on your roof, we're going to break down how each of the four main types of solar hot water systems work. Beyond these four main types, we'll also discuss some of the variations that occur within each type, to further flesh out your solar hot water options.

Direct Gain: Batch Collectors

Batch collectors, technically known as integrated collector storage (ICS), are the simplest method of solar water heating. Potable water is piped under household water pressure to the insulated solar collector, which encloses a tank or large copper tubes holding up to 40 gallons of water heated directly by the sun. This preheated water then flows through the home's existing hot water tank whenever a hot water faucet is opened inside the home.

Direct Gain: Open Loop

The second method of directly heating household water in the sun utilizes solar collectors on the roof with a large (typically 80, 120, or 250 gallon) water heater tank inside the home or basement. In this system, a differential controller activates a pump when the temperature of the solar collectors exceeds that of water in the tank. The pump circulates water from the tank to the collectors. An open loop system like this will often also pump hot water up to the solar collectors if the temperature falls below 40°F, offering a low-cost method of ensuring that the pipes on the roof do not freeze and burst. Obviously, this means losing heat from the tank, and so climates with many winter nights approaching freezing will make this an ineffective system.

Closed-Loop Heat Exchange

The first method of worry-free avoidance of freeze damage is with a pressurized closed-loop heat exchange system. The closed loop refers to a separate line of piping containing antifreeze solution (food-grade propylene glycol and water) that continuously runs from the solar collectors through the home to a specially designed water heater tank (or supplementary tank with external heat exchanger). After leaving the tank, the glycol-water mixture returns to the collectors to be reheated. The closed-loop fluid is moved by a pump activated by a differential controller that turns on when the solar collectors are a certain number of degrees hotter than the water in the tank (this can be a variable setting, depending on the controller). Since glycol can remain fluid at

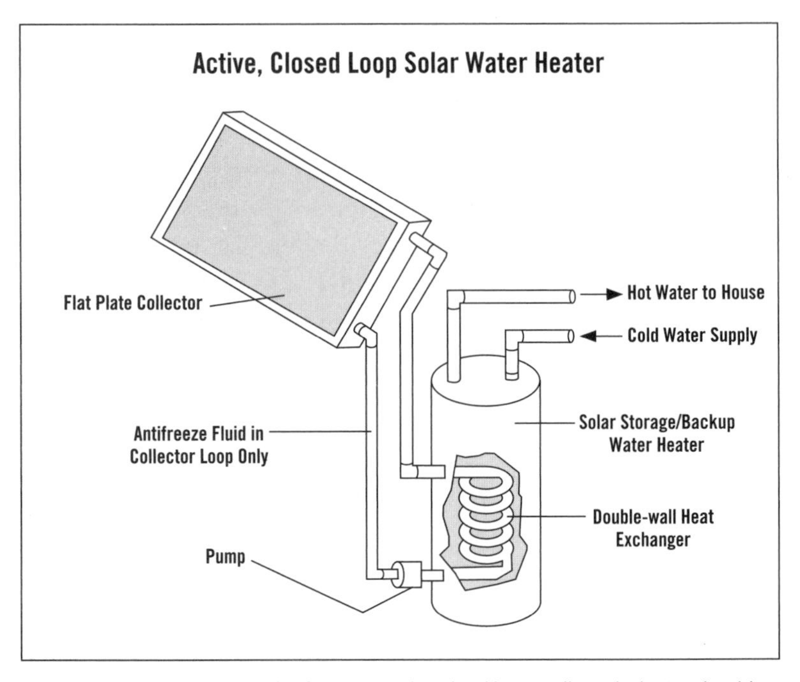

Figure 6.4. By using a fluid with a low freezing point in a closed loop to collect solar heat, a closed-loop solar water heating system can be installed in most climates of the world and effectively heat water without worries about equipment failure.

temperatures down to -50°F, this allows for effective solar water heating even in wintertime for most of the inhabited regions of the world. One problem encountered with these pressurized systems is what happens when the water tank reaches a high temperature—usually about 175°F. The pump stops circulation so the tank won't overheat, but then the fluid in the solar collectors can get quite hot as it sits in the sunshine. These systems incorporate expansion tanks and pressure-relief valves to prevent overheating. High-quality digital controllers can also start a pump up at night for a quick circulation to cool down the tank.

Drainback Heat Exchange

Equally effective in avoiding freeze and over heating damage is the drainback heat exchange system. Instead of using a pressurized fluid that could potentially become overheated and hence overpressurized,

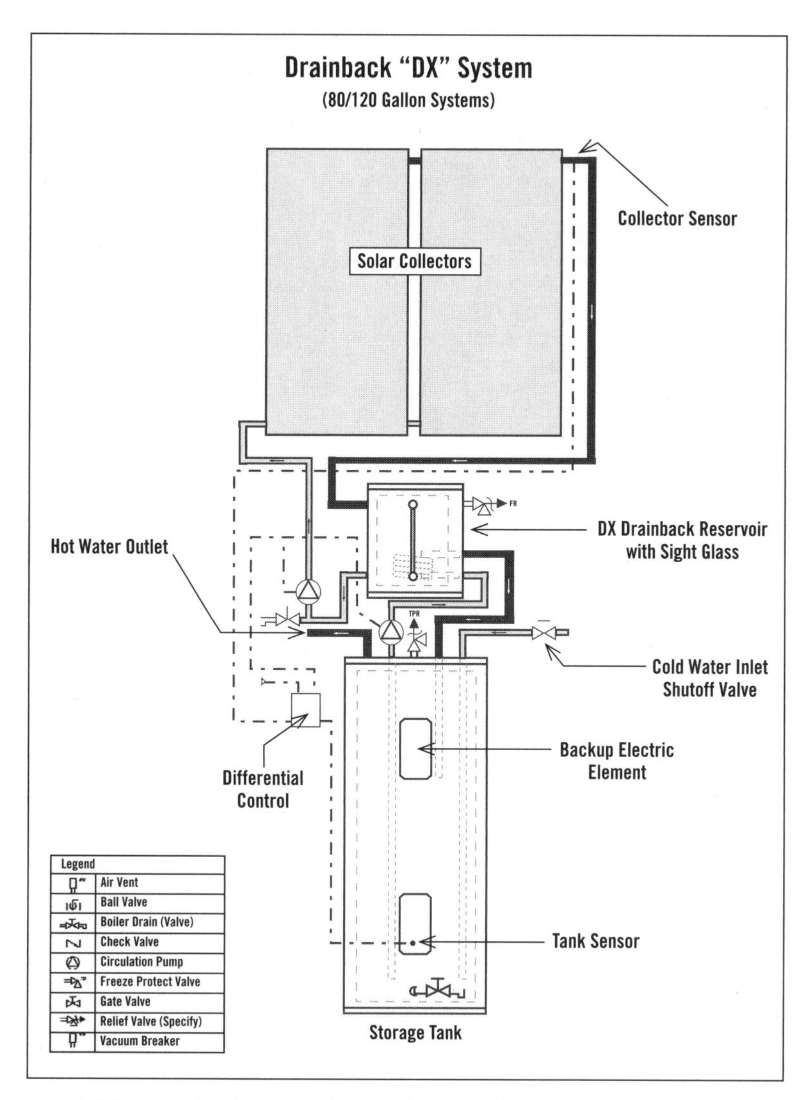

Figure 6.5. Another method to avoid freezing pipes is to use a differential controller that drains out the solar collector system when the panels are colder than the hot water tank.

the heat exchange fluid is isolated in a closed loop that empties out of the collectors and pipes into a secondary reservoir tank within the home when there's no solar heat to be gained. A pump run by a controller forces the fluid to circulate in the same manner as a pressurized system. Drainback systems should never overheat, but they can

Figure 6.6. The solar hot water heating panels are generally quite inconspicuous once installed, and studies have shown that energy-saving equipment has a positive impact on home prices. LEFT PHOTO COURTESY OF HONEY ELECTRIC SOLAR, INC. RIGHT PHOTO COURTESY OF SOUTHERN ENERGY MANAGEMENT

only be installed where the pipes to and from the collectors can run downhill all the way to the drainback tank so water won't get caught and freeze.

Solar Pool Heating

Homes with pools make excellent candidates for solar heating in the form of a solar pool heater. For those heating their pools with conventional fossil fuels, often with monthly bills in the hundreds of dollars, a solar pool heater will pay for itself rapidly. For those without any kind of pool heating system, a solar pool heater will extend the season for swimming, making the pool more of an asset to residents. These systems are efficient, straightforward to install, durable, and low-maintenance.

Solar pool heaters are nearly always simpler than their residential solar water-heater relatives. While copper-glazed pool heaters with heat exchangers do exist, the most common pool heaters are systems where the pool water is heated directly by being pumped into black UV-protected polypropylene panels mounted

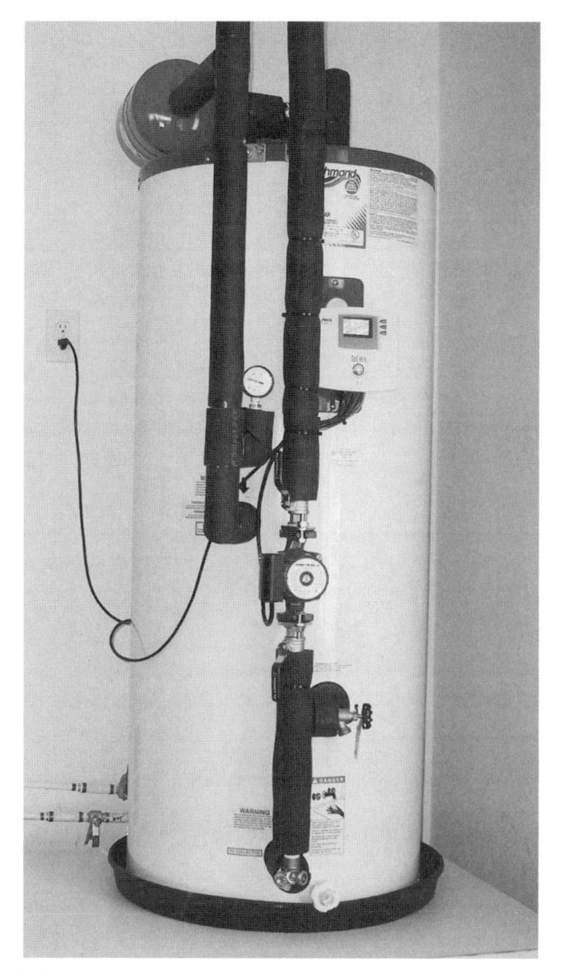

Figure 6.7. Many solar hot water installations use a single tank, usually 80 or 120 gallons for a standard household, that combines water storage with heat-exchange coils in an insulated shell. COURTESY OF HONEY ELECTRIC SOLAR, INC.

on a south- or west-facing roof or on a platform near the pool. Unlike residential solar hot water systems, these solar pool heaters are unglazed and uninsulated. Pool water is heated directly as it passes through the small diameter polypropylene tubes that are integrated into the panels, which are typically four feet by eight to twelve feet. Because of the lack of glazing that would reflect some of the sunlight away, solar pool heaters can be amazingly efficient when they are working at their prime. Efficiencies of

Table 6.1 Breakdown of Differences, Cost of Water Heater Types

	Freeze Protection	Cost	Pros	Cons
Direct Gain: Batch	Limited; works best in zone 8 or above	$3,000–$5,000 before credits/rebates	Simple and inexpensive	Adds substantial weight to roof
Direct Gain: Open Loop	Limited; works best in zone 8 or above	$3,000–$7,000 before credits/rebates	Lessens roof load compared to batch	Somewhat more complicated and expensive
Heat Exchange: Drainback	Yes; can take temps down to -50°F	$6,000–$9,000 before credits/rebates	Less prone to overheating; uses only water	Requires trickier plumbing to install than a closed loop because pipes must all drain downhill
Heat Exchange: Closed Loop	Yes; can take temps down to -50°F	$6,000–$9,000 before credits/rebates	Plumbing is easier to install and more universally applicable than drainback systems	Possibility of overheating in summertime, which can result in loss of fluid due to pressure buildup

80 percent are common on sunny days when the air temperature is above the pool temperature. In spring and fall when daytime temperatures are below the temperature of the pool, the efficiency starts to fall off. Plastic-paneled solar pool heaters cannot be used effectively when the outside temperature is freezing, so they take the winter off except in the warmest of climes.

The typical solar pool heater is similar in design to the drainback solar hot water system described earlier in this chapter (see Figure 6.8). Pools already have pumps for filtering water, and in most cases this pump can be replumbed to also send the water through the collectors. A differential controller measures the temperature of the water entering and exiting the solar panels; when there's no heat gain, the system turns off and the pool water "drains back" into the pool. This simple switch not only makes sure that the system only functions when there's meaningful heat to be gained, but it

Figure 6.8. Solar pool heaters are exceptionally effective at using solar energy to dramatically reduce fossil fuel use and high utility bills. COURTESY OF HONEY ELECTRIC SOLAR, INC.

also provides freeze protection (and prevents potential pipe bursting) when nighttime temperatures drop below freezing. Even when the panels are bypassed, the pool pump continues to send water through the filter to maintain water quality.

Keeping Your Pool Warm and Sizing the System

A general rule for how much collector space you'll need depends on the surface area of your pool more than its volume. That's because up to 70 percent of the heat loss from your pool water happens via evaporation from the surface of the pool, and another 20 percent from radiation to the sky. So before we can look at the amount of collector space you need for your solar pool heater, it's vital to look at how to cover your pool up when it's not in use.

Pool covers run the gamut from simple layers of what is essentially UV-protected bubble wrap to more durable (and expensive) vinyl covers. The bubble covers are also sometimes called solar pool covers, and they do the best job of eliminating evaporation and reducing radiation losses while still allowing for substantial solar heat gain. The drawback is the shorter life span (about five years) compared with vinyl covers, which are better suited to withstanding the harsh rays of the sun and corrosive pool chemicals. Vinyl covers are longer lasting (about ten years) and come either uninsulated or insulated. Vinyl covers block about 25 to 40 percent of potential direct solar gain. Any of these covers can be mounted on a reel for ease of use, and if you're willing to shell out the dough, you can buy a motorized wheel that will do the cranking for you. For an insulated vinyl cover with a motorized wheel, you might end up shelling out a thousand dollars or more, but the amount of heat loss prevented is staggering, and even an expensive cover will easily pay for itself in a year or two. Besides the energy savings, pool covers reduce needed make-up water by 30 to 50 percent, reduce chemical consumption by 35 to 60 percent, and lessen cleanup efforts because debris is kept out of the pool. Another option for keeping your pool insulated is with a liquid pool cover, a safe and biodegradable product that

Figure 6.9. Solar pool heaters are much more effective when used in combination with a pool cover, whether liquid or permanent. COURTESY SMC.

is added regularly and sits on the top of your pool water, retaining heat.

Another effective way to reduce heat loss in windy areas is with a windbreak of planted bushes on the side of your pool where prevailing winds come from. For this to be effective, the line of bushes should be close enough to the pool (within ten feet, and at least head-tall once grown) to block the wind but not shade the pool.

Once you've figured out your cover situation, a general rule of thumb for sizing pool heaters is to have a collector surface of roughly half the surface area of your pool in southern climates, and two-thirds in northern climates. This assumes a good south-facing roof with plenty of solar access for the pool *and* the panels during heating months (spring, summer, and fall)—and of course a good pool cover. In less than ideal situations such as a west-facing roof, you'll want to increase the collection area if possible.

Figure 6.10. Solar pool heaters can also be installed on the ground near the pool if the roof does not offer a good option. Installation often is completed in a day. COURTESY OF HONEY ELECTRIC SOLAR, INC.

Costs and Payback

A typical solar pool heater will cost in the $3,000–$5,000 range for an installation. With pool heating costs often at least several hundred dollars (and not uncommonly above $1,000) a month, the return on investment can be dramatic—even before state and local incentives. These systems are simple and easy to install, with experienced technicians often able to complete an installation in a day. With low maintenance costs and effective technology, all else being equal, solar pool heaters are able to pay for themselves quicker than any other type of solar technology. And of course they continue to save you money long after installation costs are recouped, all the while protecting the planet and making for a very pleasant swim.

Resources

Homepower: http://www.homepower.com
Liquid Pool Cover: http://www.liquidpoolcovers.com

CHAPTER 7

Solar Heating

Since heating is a major consumer of fossil fuels for most homeowners (often *the* major consumer), and sunshine is good at heating things up, it shouldn't be surprising that solar space heating is piquing people's interest. Whether by adding windows to the south side of a building, installing a solar air heater, or integrating a radiant floor heating system with a solar water heater, you can cut your carbon output and heating bills at the same time.

Solar heating is unique in that it is only needed part of the year—in the wintertime. Obviously, the longer your winter, the more benefit you'll get out of any potential solar heating options. Like other solar options, solar heating is intermittent and should

Q. Does solar heating really work?

A. Yes, definitely, but only when the sun is out! This means a backup heat source will be necessary for most spaces, especially residences, that need heat at night or during cloudy days. The more insulated your space is, the less you have to rely on backup sources.

Q. Does solar heating provide meaningful heat all winter?

A. Solar heating tends to be most effective in early winter (October and November) and late winter (February and March). Not only are days short in the depths of winter, but the solar energy needed to provide meaningful heat to the solar collection system must also overcome colder outside temperatures in December and January. This means that a supplemental heating system is required for most climates, although its use will be curtailed substantially once the solar heating system is installed.

Q. Why aren't solar air heaters more common?

A. Honestly, we don't know. They are a relatively inexpensive and low-tech way to retrofit solar energy into buildings.

always be considered as a supplemental means of heating your home, garage, or office. Also, because the sun is lower in the sky during winter, the amount of potential obstructions—such as trees, buildings, and power lines—increases. The good news is that solar heating options are unique in that they can often be installed on south or west-facing walls rather than just on roofs, so if your roof is already full of PV modules and hot water panels, there's a whole other area of your home just waiting for more solar!

If you are reading this in the wintertime, the simplest way to get an idea of the suitability of your south-facing walls or roof for solar heating is to just take a look (or even better, a picture) of where you're thinking about getting a heater installed at 10:00 A.M., noon, and 2:00 P.M. The site needs to be in complete sun at these three times throughout the winter months. Don't make any assumptions! This is why taking pictures is so helpful, as memory can be overly optimistic. The closer your observations are to the winter solstice (December 21), the more reliable the information, as this is when the sun is lowest in the sky. We discussed evaluating your home's solar potential in Chapter 2, but because solar heating equipment is different from solar hot water and solar electricity in that its use is limited to when the sun is lower in the sky, it's important to keep the position of the sun in wintertime in mind.

The Four Types of Solar Space Heating

If you've got winter sun available, there are two basic systems for accessing it for heating purposes, and two primary types within each system. The first system is called **passive solar heating**, and is accomplished by allowing winter sun in through south-facing windows, and keeping the sun out in the summer through some type of overhang. This is typically something that is incorporated into the initial design of the building, but passive solar gain can also be done as a retrofit. When passive solar heating is built or retrofitted into south-facing walls of the main living area, we refer to it as **building-integrated** passive solar heating.

Figure 7.1. Existing homes can access solar heating in a variety of ways, including passive solar. This masonry home has opted to add a sunspace, or attached passive solar greenhouse, onto its south side to collect solar energy. Climates with warmer summers need to ensure such a space won't overheat. PHOTO BY W. L. TARBERT, WIKIPEDIA.

The second type of passive solar heating that has received something of a bad reputation is a **passive solar addition**, usually referred to as a **sunspace** or **passive solar greenhouse,** on the south side of the home. We will use the term *sunspace* to refer to additions that employ sloped glazing; for passive solar additions that employ vertical rather than sloped glazing, we will use the term *solar porch.* The key distinction for both is that they can be physically blocked off from the main living area of the home.

The second two systems are **active solar heating systems**, meaning they use a fan or pump. The simplest of the two types of active solar heating systems is a **solar air heater**. Relatively rare—but effective in colder, sunny climates—is the second type, **solar radiant heat** (also referred to as a **hydronic solar radiator**), which typically combines an expanded solar hot water system with radiant piping inside the home, either wall-mounted or integrated in the floor.

Solar heating has long been the province of the do-it-yourselfer, and premanufactured models and knowledgeable builders have only

Figure 7.2. Solar heating systems are unique in that they can be placed on the south or west side of a building, rather than on the roof. In fact, it is simpler and more effective to install them there. Unlike passive solar heating, active solar heaters only operate when the sun is out, and thus prevent heat loss through large windows when there is no solar gain to be had. PHOTO COURTESY YOUR SOLAR HOME, INC.

begun to reappear in the last few years (after a long hiatus caused by the removal of tax incentives and the decline in price of fossil fuels in the 1980s). Since solar heating options have only recently made it into the mainstream, finding a reputable installer can be a challenge. (Check Chapter 4, "Getting Ready for the Installation," for more details on finding reputable solar installers.)

What's Best for Your Climate

Heating requirements, unlike hot water or electric needs, vary greatly from region to region—from little or none in south Florida to an almost continuous need in colder parts of the continent. Another crucial distinction for solar heating is that the efficiency of the system can vary greatly depending upon the amount of cloudy weather. Essentially, every location has its own heating demand requirements and its own potential solar heating supply. Different solar heating systems perform better in different locations, so some background legwork on the homeowner's part will ensure that an inappropriate system is not installed by an overzealous solar installer. It's also important to remember that solar heating systems tend to work best in late fall/early winter and late winter/early spring, when days are a little longer and temperature differentials between inside and outside are less.

Since climate and solar variability means that different parts of the country should consider different approaches to solar heating, we hope to simplify your decision process by offering rough guidelines for choosing an appropriate option based on those two factors. These are guidelines, and they come with the caveat that the different methods of solar heating provide additional advantages and limitations other than how effectively they heat up your home. But given several decades of modern solar heating history and experimentation, we feel that a set of practical guidelines for the average homeowner is sorely needed. Before we get to these guidelines, however, we will clarify a few more distinctions.

Fundamental Differences between Passive and Active Solar Heating Options

Passive and active solar heating options both function by converting sunlight into heat for the home, but beyond this basic fact function very differently. Cost is the most obvious factor, but there are also fundamental differences in the amount of daylighting, personal involvement, ease of retrofit, aesthetic concerns, and more that depend upon the individual preferences of the homeowner.

Passive solar is incorporated into the design of the home. Despite its name, passive solar heating functions much more efficiently with the *active involvement* of the home's resident. Passive solar heating requires replacing well-insulated and well-sealed wall space (typically around R-15 insulation value) with some type of glazing (typically around R-1.5 to R-3). To prevent additional heat loss when there is no solar gain to be had, an insulated curtain or shutter closes up the windows and increases the R-value to something like R-5 to R-10—still below typical wall space but much improved over simple glazing, even double-paned.

The basic idea with passive solar heating is that when there is solar heat to be gained, the shutters are opened; when the sun goes away, they're closed. Successful passive solar requires manual manipulation of the solar heating equipment. This manipulation can either be via a temperature-controlled fan or curtains, or more commonly by the physical presence of an interested person. Without thermostatically controlled "automatic" functioning, in parts of the country where solar gain can come or go frequently because of cloudy weather (or if the occupants of the home have schedules that leave the shutters open when there's no heating occurring), the windows can have a negative effect on the amount of heat entering the home compared with insulated wall space. Likewise, in the spring or fall excessive amounts of heat can be added to the living space through the windows if the shutters remain open when it is sunny outside.

Since passive solar is integrated into the building itself, retrofitting passive solar heating into an existing home can require major construction work. In a common retrofit, some or all of the south-facing wall will be torn out (with the roof supported during the process) and replaced with windows. Alternatively, a passive solar porch or sunspace can be constructed onto the south-facing wall. In most situations an overhang, awning, or trellis that blocks the sun in late spring through early fall helps keep unwanted heat out of the living space.

While this may sound like a lot of trouble and expense to go through, it may be worthwhile if there are other advantages that appeal to the homeowner such as daylighting, additional living space, aesthetic concerns/improvements, or access to a pleasant view. For a traditional stick-built home, opening up a wall and reworking it are common building practices that can be accomplished by a competent professional, although reworking a masonry building will be much more involved. A sunspace or solar porch is an addition onto the existing home, and so doesn't require any structural changes to the home. But these additions present their own challenges of moving any captured heat into the home itself and potentially providing unwanted heat during summer. The permanence of passive solar retrofits, whether adding south-facing windows or an attached sunspace or solar porch, means that once the work is completed, the home will have access to solar heating for the rest of its existence. This provides a measure of longevity above an active solar heating system.

To complete the misnomer, active solar heating allows the *passive involvement* of the homeowner. Heat is transferred by moving air or water only when the sun is shining and meaningful heat can be gained. The fluid's movement is regulated either by thermostat or electrical production of a solar electric panel that powers the fan or pump. Active solar heating is accomplished with solar collectors, typically large (four by eight or four by twelve foot) black rectangular boxes that will be mounted on either your wall or roof (or also potentially using evacuated tubes, discussed in Chapter 6). Because active solar heating employs separate collectors, it does not involve major construction work as a passive solar retrofit would. The collectors are active, meaning they rely on moving parts—either a fan for air systems or a pump for radiant (hydronic) systems. These can stop working, either due to blockage or breakage—and the collectors themselves can become degraded over time. Instead of a permanent retrofitting of your home, active solar heating systems should be considered an addition on par with a new roof or good paint job, something that will last around twenty to thirty years. It's important to note that solar heating panels mounted on a wall and under a small overhang will be kept out of wet weather and summer sun, and will correspondingly last much longer than if mounted on the roof (as well as not adding potential leaks to your roof).

One Last Difference: Radiant Heat versus Forced Air

The last major difference in solar heating options is *how* they provide heat to your home. Heat moves in three ways—directly through conduction, by expanding and moving a fluid such as air or water through convection, or electromagnetically from object to object via radiation. This is important because the three methods of heat transfer directly relate to our perceptions of comfort—which is, after all, what we're trying to maintain. Generally speaking, we *Homo sapiens* find heat to be most comfortable when it is transferred radiantly. The reasons are complex, and involve mean radiant temperature as well as a general distrust of moving air, which we rightly fear stirs up things like mites and dust that should just as well be left where they are.

Radiant heating is provided directly from the sun itself, so direct gain from passive solar heating is the simplest method of achieving this, and there's no disputing the warmth and pleasure provided by sitting in the sun on a chilly winter day. The other option for radiant heating is a hydronic solar radiator. Rather than heating the inside of the home directly through the sun's radiation, a solar radiator heats up water or glycol in the sun and then distributes this heat inside the home through piping, either in the floor or through a metal radiator inside the room, often wall-mounted.

The other two solar heating systems (passive solar additions and solar air heaters) require the forced movement of air to effectively distribute the heat captured. A solar air heater circulates air through an intake or outtake in the wall or ceiling by means of a fan. The heat captured in an attached sunspace or solar porch should also be moved by a fan in order to reach the interior of the home in significant quantities.

Determining Your Climate Zone: Heating Degree Days and Winter Insolation

Firsthand knowledge of where you live lets you know if you have much sun in the winter or how cold it gets, but for general purposes of determining the type of climate you live in and the corresponding best method of utilizing solar power for some of your heating needs, you'll need to determine two things. First you need to know how much heat your home needs, based on your climate. And second, you'll need to know the potential supply of sunshine during winter for your area. Although this section requires a little legwork for the homeowner, we strongly recommend you take the time to better understand your climate so you can engage knowledgeably with any potential installer. Solar heating is often misunderstood, and the number of poorly done installations we've witnessed is testament to the fact that professional expertise in this area is sometimes lacking.

Your home's heating needs are most easily determined by finding out how many **heating degree days (HDDs)** your location has, determined by historical data. On any given day, HDDs are determined by taking the average temperature for a twenty-four-hour period and subtracting it from 65°F. It's been determined that when the outside temperature is lower than 65°F, additional heat is required to keep the inside of the house near what is considered a comfortable 70°F. For example, a day that averages 40°F gives 25 HDDs (65 – 40 = 25). To get a tally for the winter, each day for the year is tallied up, with a zero value for any day that averages above 65°F. This winter total provides a quick determination of your climate's heating needs (see "Resources" for info on determining your locale's HDDs).

Once you know the heating demands of your climate, the other piece of the puzzle you'll need to determine is your potential supply of sunshine. This measurement is winter insolation, or the amount of solar energy that reaches your location each day (typically represented in kWh/m^2) and based on historical data. It's important to remember that we only want data for winter, and to disregard annual averages. Not only are the days shorter in winter, but some climates—especially places near the ocean like much of the west coast—have sunny summers but cloudy winters.

The National Renewable Energy Laboratory has accumulated insolation data for points all over the United States (and to a lesser extent the world) that are broken down by month. This information is available online through their PV Watts calculator (search for "PV Watts Version 1"). An important caveat for the use of the solar heating breakdown in the following section is that we are assuming insolation numbers on the default tilt angle that PV Watts provides for each location, which is equal to latitude. These insolation numbers will be slightly higher (10 to 15 percent) than for the vertical installation of solar heating equipment that is most common, but it allows a rough comparison among solar heating types and various climates. We encourage you to play around with the PV Watts site and fine tune your understanding of how changes in tilt angle (for example, an installation on your south-facing roof versus south-facing wall) would potentially affect the amount of solar energy your site receives. Generally speaking, *for solar heating equipment only* the ideal tilt angle will be your latitude plus 12 degrees (12 degrees is about half the movement of the sun from the fall and spring equinox to the winter solstice, which roughly corresponds with peak heating loads; see the discussion of tilt angle in Chapter 2, "What's Appropriate for your Site").

Making the Choice: Determining the Appropriate Solar Heating Option for Your Home

The basic passive solar heating principle of having a majority of the home's windows on the south-facing side of the home should be employed by *every* new home with significant heating requirements (that is, HDDs greater than 1,000). Passive solar heating is extremely simple yet very effective in providing substantial amounts of heat, and since homes will have a certain amount of wall space devoted to windows anyway, it makes sense to design the home so that these are on the south-facing side.

Every home is different, and how it should be heated also depends on factors such as the habits of its residents. Will the inhabitants be willing *and able* (i.e., physically present at the appropriate times) to assume responsibility for the moveable insulation that is typically a fundamental aspect of effective passive solar heating? We will assume that they will, but if they won't be willing and able, it may make sense to use a more automated solar heating system—either employing thermostatically controlled curtains or by installing an active radiant or air system.

With this caveat, we'll look at some broad parameters for what type of solar heating is appropriate for various climates.

Majority of Heat from Building-Integrated Passive Solar. If, according to PV Watts, your location averages 4 kWh/m^2 or more per day of insolation for the winter months (i.e., those months when heating is needed at least half of the days of the month), and requires at least 2,000 HDD for the entire season, then a standard passive solar design for your home will work very effectively. For locations with less than 2,000 HDD but at least 4 kWh/m^2 of insolation, then simply having the majority of the home's windows facing south should provide most of its heating needs.

Majority of Heat from Active Solar (Air or Radiant). Active solar heating is an effective solar strategy for locations that receive an average of at least 3 kWh/m^2 during winter months according to PV Watts, and have heating requirements of at least 3,000 HDD. Below these numbers, the cost and maintenance of the system may make it prohibitively expensive compared to other effective sources of alternative heating such as a geothermal system (which also helps with cooling) or masonry heaters and woodstoves. The reason to use active solar heating in these circumstances is because it operates automatically only when the sun is shining and meaningful heat is being provided. Active solar heating eliminates the heat loss from a large bank of windows when the day is cold and cloudy.

Active solar heating may be chosen in locations that receive more than 4 kWh/m^2 instead of passive solar for the following reasons: inability or unwillingness of occupants to operate moveable insulation; need for retrofits where it is easier to add something to the wall or roof than modify the structure; and homes where the roof provides solar access in winter but the south-facing wall is too shaded.

Supplemental Heat from a Passive Solar Addition. For locations that average less than 4 kWh/m^2 in the winter but still have substantial heat loads (HDDs greater than 2,000), then an attached passive solar addition can make an effective supplemental solar heating source on sunny days, as well as providing a wonderful hangout spot for sun-starved inhabitants. It is very important to note that sunspaces (again, those additions with sloped glazing) have a tendency to overheat during the warmer months of the year because shading the glazing is usually quite difficult (obviously, this varies for each location; a solar site analysis of the sun's path in summer might show a well-shaded location). Because of this concern of overheating in summer, if your location has substantial cooling loads* (or cooling degree days [CDD] greater than 1,000), then you should limit your passive solar addition to a solar porch (unless the location is heavily shaded). Solar porches that employ opening windows can make great hangout spots in winter and summer. It may also be relatively easy to retrofit an existing south-facing porch into a solar porch.

Getting Down to Business: The Nitty-Gritty of Your Solar Heating Installation

Once you've double-checked your wintertime solar access and determined the best solar heating option for your home, family, and budget, it's time to look at the specifics of each heating system so you know what you're talking about when the installers or builders come to give an estimate. Unlike solar water heating or photovoltaics, solar heating options are more varied and nuanced, and understanding the basic principles of the system you are considering will go a long way toward helping you weigh the various choices before a successful installation.

Sizing Your Solar Heating System

Whatever system you choose, you'll need to have an idea of the necessary size. Assuming your location meets the criteria for solar

---

*Calculated in the same manner as HDD, and available from the same Web sites. A summer day's **cooling degree days (CDDs)** value is the average temperature for the day minus 65°F. For the season, CDDs are tallied for every day that averages above 65°F.

Table 7.1. A Quick Breakdown of Solar Heating Options Based on Climate and Sun-Hours			
Majority of Heat from . . .	**If . . .**	**Benefits**	**Concerns**
Building-Integrated Passive Solar	HDD* > 2,000 and sun hours > 4	Daylighting, permanence, view	Requires inhabitants' involvement unless triple-paned windows or automatic shutter are used, major construction to home
Active Solar Heating (Air or Radiant)	HDD > 3,000 and sun hours > 3	Operates automatically, easier retrofit, potentially roof-mounted	More limited lifespan (twenty to thirty years), roof-mounting raises concerns of roof age and leaks
Passive Solar Addition	HDD > 2,000 and sun hours < 4; for sunspaces CDD < 1,000	Functions well in cloudier climates, easier retrofit, solar porches with opening windows can make good hangout spots in summer	Overheating in summer for sunspaces, usually requires occupants' active participation

heating needs outlined in the "Making the Choice" section above, a general rule of thumb is that the square footage of the solar collection area (windows or panels) should be roughly equal to 7 to 10 percent of floor area to be heated, assuming no **thermal storage**[*] (in which case, the percentage could be higher). For example, a twenty by twenty foot room (400 ft^2 total area) could be heated by twenty-eight to forty square feet of solar collection space. So a typical four by eight foot solar air heater, or three by five foot windows (two of them), would make an effective solar heating option, assuming an open solar window. Milder climates should stay near the 7 percent number; colder climates should aim for 10 percent.

[*]Thermal storage refers to dense building materials such as stone, concrete, or tile within the building that are capable of storing heat during solar gain and releasing it in the evening. For more detail on this topic, see *The Carbon-Free Home*.

South-Facing Windows, Overhangs, and Insulated Curtains

Passive solar heating depends on getting the sun in when you want it (winter) and keeping it out when you don't (summer). Fortunately, the sun is very accommodating in accomplishing this, shining on the south side of our buildings in winter, and east and west sides in summer. Utilizing this moving heat source effectively involves a combination of the correct window technology on south-facing walls and, in most cases, some type of overhang, awning, or trellis to block the sun when it's higher in the sky. For large amounts of passive solar, it should also include some type of insulated curtain or shutter to bring the insulation value of the window area closer to a typical wall (about R-15) when no meaningful heat gain is occurring.

Unfortunately, there's no boilerplate design that can be applied to different regions of the country for passive solar design. Each location has its own peculiarities and microclimate, and combinations of different windows with different overhangs will produce a range of potential heating options. The biggest variables are, first, if you need overhangs at all and, second, how and where to place them and at what time during the spring and fall they block the solar heat gain. Any potential builder that you are considering hiring for such a major undertaking should have experience designing such a system *for your particular area and climate*. Ideally, there would be an example of a functioning passive solar home nearby where you could talk to the owners and see how well it's working (Does it get too hot in spring or fall? Does it get too cold at night? What about issues of privacy, daylighting, and so on? Because of the nuanced nature of passive solar design, it is the most likely solar heating system to be installed incorrectly, and there's no doubt that a poorly executed passive solar design will result in both excess heat loss in winter and excess heat gain in summer. One good Web site that can be very helpful in understanding overhangs and windows in passive solar design is "Sustainable by Design" (see "Resources"). Always keep in mind that any window and overhang combination is working within the limits of your particular solar window and the sun and shading pattern it produces.

Windows

Windows are much more than just panes of glass, even if they're fixed in place and don't open. Over the last several decades, major advances have been made in making windows more insulating and less prone to suck heat out of our homes. This is a big deal, because windows are typically major sources of heat transfer, not just through draftiness of insufficiently closed sashes, but because they are thin barriers that barely separate our inside from the cold or hot outside.

Advanced window technology means that properly planned passive solar design will perform better. But it also means that it's easier to mess up by using an inappropriate window system. Being knowledgeable in this area is crucial, especially if you are dealing with a traditional builder with little expertise in passive solar design.

Purchasing windows has become almost as complicated as buying a new car, with the number of options and varieties seeming to grow by the day. Going out window shopping (literally) might reveal choices ranging from single-paned, double-paned, triple-paned, krypton-filled, vacuum-sealed, low-e—the choices seem to go on and on. Fortunately, the characteristics of all these different options, especially as they relate to passive solar heating, can be summarized by a few criteria. The relevant information is on the National Fenestration Rating Council (NFRC) label that should be prominently displayed on any window you're considering for purchase (see Figure 7.4). To effectively use windows for passive solar heating each window installed on the south side of your home *must* meet the following criteria:

- Have a solar heat gain coefficient (SHGC) greater than 0.60
- Have a U-factor (its insulation value, more comprehensive than R-value) of 0.35 or less
- Have a visible transmittance (VT) greater than 0.65
- Be certified by the NFRC
- Be certified by EnergyStar

While you want heat to enter your south-facing windows, any windows on the east and west sides of your home should have a low SHGC (0.40 or less) to keep out morning and afternoon sun in the summer.

Don't take anyone's word on these attributes. Check the label on the window yourself, or you could be setting yourself up for a very costly mistake. The rest of the information about the window—low-e coatings, krypton- or argon-filled—is just icing on the cake. Windows come with various guarantees and warranties; keep these, along with your receipt. Although there has been great improvement since double-paned windows were first introduced, don't take the manufacturer's word that the seal around the panes of glass won't fail and lead to a catastrophic buildup of condensation between the panes a few years down the road.

For folks who want a passive solar installation without worrying about opening and shutting an insulated curtain or shutter, your choice is either to install an automatic thermostatically controlled shutter, or you'll need to spend the big bucks and purchase triple-pane windows that should meet all of the above criteria except the U-factor listing should be less than 0.20. At this level of insulative value, the window assembly is approaching a typical wall R-value. These are the type of windows that are used effectively in *passivhaus*-style residences, living units that are capable of providing all of the needed heat using solar energy in locations as far north as Sweden and Minnesota.

Overhangs, Awnings, and Trellises

Adding windows to your home's south-facing wall is great for letting in more heat in winter, but it can also lead to overheating—especially in late summer, but also in spring while the sun is still low enough in the sky to send sunshine through your windows. To increase shading during these times of year, passive solar design usually requires some type of overhang, awning, or horizontal trellis above the windows. An **overhang** is generally considered part of the building, while an **awning** is attached separately, and a **trellis** is made of wire or thin wood and has a living vine growing on it.

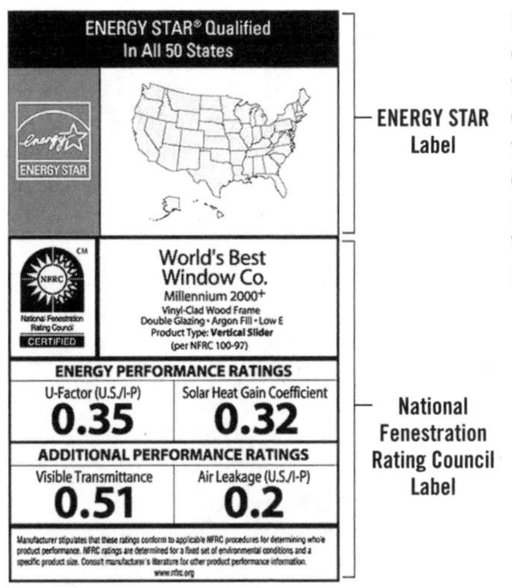

Figure 7.3. Using passive solar heating effectively requires some homework on what windows to install on the south side of your home. Among other things, installing windows with a low solar heat gain coefficient <.60 can result in little or no solar heat gain. Fortunately, every new window comes with a label from the NFRC that shows the relevant information. SOURCE: NFRC.

Generally, awnings and trellises are easier to retrofit onto an existing building. Shading in summer increases in importance with the warmth of your summers.

It's also important to know that a trellis with a deciduous vine such as a grape will provide superior shading compared with a fixed-in-place awning or overhang. This is because the heating and cooling seasons lag behind the shortest days (when the sun is lowest) and longest days (when the sun is highest) of the year. So while March 21 has the same sun path as September 21, heating needs are still required in March while they are generally not wanted in September. Deciduous plants adhere to the heating/cooling cycle rather than the solar cycle, so a deciduous vine will allow in March sunshine, while blocking it in September, for instance.

Insulated Curtains and Shutters

Extra windows on south-facing walls increase potential heat loss, as the windows are far less insulated (and potentially draftier) than framed-in wall space. Unless you are installing very high-end triple-pane windows with a U-factor of below 0.20, having moveable insulation such as an insulated shutter or curtain for your passive solar windows will greatly improve their effectiveness. Besides air leakage, heat loss through windows occurs in two different ways:

Figure 7.4. Deciduous overhangs can be a simple, and beautiful, way to shade south-facing windows in summertime. When the weather cools off and the leaves drop, the sun comes pouring in.

through radiation from the glass itself (low-e coatings help minimize, but do not eliminate, this) and through convection, as air next to the window is cooled and falls, to be replaced by additional warm air from above. Traditional curtains curtail radiation heat loss, but can actually exacerbate convection losses. This happens when air is allowed to flow between the curtain and the glass—entering from above, becoming cooled, and exiting through a gap at the bottom of the curtain and window. Due to a phenomenon commonly referred to as the **chimney effect** that results in increased air flow between two surfaces compared to what would occur with an open surface, curtains can actually accelerate heat loss despite reducing losses from radiation.

The long and short of this is that curtains and shutters are much more effective if they block air flow, especially at the top and bottom. The next thing to look for is how insulative the curtain or shutter is. This generally means several layers of material, with one of the layers being a flexible type of insulation. Good insulative curtains or shutters will also often have a radiant barrier as the last layer to the outside. Custom-made curtains can get pricey, above $\$10/ft^2$, and can add significant cost to a passive solar installation.

Sunspaces and Solar Porches

The first wave of passive solar enthusiasm in the 1970s produced a lot of attached south-facing sunspaces (sunrooms, attached greenhouses, and so on), many of which did help add heat in the wintertime, but did a much better job of it in the summertime. Lots of bad installations gave sunspaces a bad reputation (much as it did for shoddy installs of solar hot water and air heaters), a reputation that has generally survived into our current solar revival. As mentioned in the introduction, the concept of a sun space or room on the south side of the home has been around for millennia, and was the primary method of capturing solar energy by the sun-worshipping Romans (called a *heliocaminus*). As a simple, homeowner-operated solar heating option, sunspaces are relatively inexpensive compared with building-integrated passive solar heating. Installed in a proper climate and with willingness on the part of the residents, they can make a great hangout spot during sunny days—and by leaving access such as windows and doors to the space open during the day and blowing in the hot air using a fan, they can provide substantial supplemental heat. They also make a great spot for cold-sensitive plants and for starting seedlings, making sunspaces a good choice for gardening enthusiasts.

For climates where overheating in summer is a greater concern (CDDs greater than 1,000), the more expensive option of a solar porch should be considered. With openable or removable glazing, solar porches are sunny and warm in winter, and cool and breezy in summer—the perfect combination. We converted our south-facing porch into a solar porch, and derive a great deal of supplemental heat from it. The glazing is removed in April and replaced with screens, and deciduous overhangs keep the sun outside producing yummy grapes and hardy kiwis rather than heating up our favorite hangout spot.

Active Solar Heating Options: Air or Liquid?

If your climate and preferences suggest that an active solar heating option is the way to go, the next decision is whether you want to pursue a solar air heater or a solar radiator (aka solar hydronic heater, or solar space heating). There are a variety of factors to look at that will help you decide. These are the differences in size, cost, and ease of installation; position on wall or roof; radiant or forced air heat; and potential tie-in with existing hot water or heating systems.

Wall versus Roof Mount

There are exceptions, but all else being equal, solar air heaters are easier to install on walls, while solar radiators can be installed on walls or roofs—or even out in the yard or on a separate outbuilding. Solar air heaters perform much better if they have an "in" air register near the floor and "out" register near the ceiling. Not only does this work in tandem with the natural thermosiphon within the heater (heat rises), it also promotes air circulation, and hence temperature mixing, within the room. When air registers are installed on a ceiling, as is more likely with the collectors on the roof, an air heater tends to produce a stratified layer of heat near the ceiling—not much use to the average-sized human. This stratification can be compensated for by increasing the size and air flow of the blower or with the aid of a ceiling fan, but at additional cost and energy usage.

Your home's solar window is likely to be better on your south-facing roof than your south-facing wall, simply because it's higher up and more out of the way of potential obstructions. Since liquid can be moved through inch-wide pipes instead of a six- to twelve-inch-diameter insulated duct, installing a hydronic system rather than an air system on roofs generally make more sense. This makes any potential intrusions into your roof, ceiling, and walls substantially less, and means heat can be moved greater distances and through smaller spaces (especially stud-framed walls).

Cost and Ease of Installation

Solar air heaters are simpler devices than solar radiators, meaning they cost less from the get-go. An air leak is much less potentially destructive than a liquid leak, and air is under much less pressure, making preventing heat (not roof) leaks easier. Also, since solar air heaters generally get mounted on walls rather than roofs, a small-scale installation is less invasive and less dangerous—making the amount of labor less, and the job cheaper. The labor is also more straightforward and less skilled, meaning a general handyman can successfully install a solar air heater, while a solar radiator requires skilled plumbing work. Solar radiators work well with a radiant heat floor system—but make no mistake about it, radiant floors are costly installations.

Since moving air is more difficult than moving liquid, solar air heaters tend to make sense for single-room applications, while solar radiators make sense for larger, multiple-room systems. Solar radiators have fixed costs—such as piping to and from the panels, the pressure tank, and getting workers on the roof—and these fixed costs tend to be less onerous the larger the system is, bringing the cost down for larger systems. Of course, a larger system only makes sense if you actually need the heat. It should be noted that tying in a solar air heating system with the existing furnace is possible, although installations are uncommon. Such large air heating systems probably suffer a lower efficiency because of the additional energy needed to move so much air.

Hot Water Tie-In

Radiant heat systems have often been tied into the hot water supply of residences, and it may be possible to catch two birds with one net if you are considering both solar hot water and solar heating. Combining the two systems could save money, as some of the fixed costs mentioned previously can be combined for both. Also, solar water heating systems are more specifically outlined in the federal tax credits than other solar heating options (see "The Cost of Solar Heating Systems" in Chapter 3). The drawback is the added complexity of a combined system, and the failure potential of both heat and hot water occurring simultaneously.

Considering all of these differences carefully before talking to contractors and getting estimates will greatly improve the chances of having a productive conversation about the right system to install for the cost. While operation with an active system will be straightforward and relatively carefree once installed, doing the legwork of figuring out the best option for your home will keep you from buying a system you may not want or gives you poor results.

Evaluating Equipment and Installers for Solar Heating Systems

Passive solar and sunspace installations are work that can be carried out by any competent carpenter or builder. Using only insured and bonded contractors is a great idea. Seeing evidence of previous successful installations also goes a long way toward alleviating concerns. Once you've found a competent installer, the big concern is purchasing the correct type of window, so make sure the windows used in your passive solar installation meet the criteria we outlined in the "Windows" section on page 119.

Solar air heaters are relatively easy to construct—being basically just a glazed, insulated black box—so manufacturers tend to come and go. Obtaining a quality panel is probably the biggest hurdle of a successful solar air heater installation. The best way to ensure a quality product is to make sure it is rated by the SRCC (Solar Rating and Certification Corporation). Solar air heaters are fairly straightforward to install, especially on south-facing walls (roof installations tend to be substantially more complex). The same considerations that apply to passive solar and sunspace installations apply to solar air heaters as well.

Solar radiators are much more complicated, but because they are so similar to solar hot water installations, any competent NABCEP-certified solar hot water contractor should be able to figure out the installation. Follow the steps outlined in Chapter 4, "Getting Ready for the Installation." Solar radiator equipment is identical to solar hot water equipment, and should be SRCC rated.

Resources

Your location's HDDs can often be found in the weather section of your local newspaper or local news Web sites. Three Web sites that offer free and accurate HDD information (each of which have their own quirks) are:

National Climatic Data Center (NCDC): http://lwf.ncdc.noaa.gov/oa/documentlibrary/hcs/hcs.html

Weather Underground (also available in Celsius): http://www.degreedays.net

Weather Data Depot: http://www.weatherdatadepot.com

Sustainable by Design Window-Overhang Toolkit: http://susdesign.com/overhang

EnergyStar Window Information: Search for "EnergyStar Windows"

National Fenestration Rating Council: http://www.nfrc.org

Insulated Curtains: http://www.cozycurtains.com

CHAPTER 8

Everything Else under the Sun

To supplement your new solar electric, hot water, and heating systems, there are many varieties of solar gadgets and gizmos that can accomplish a wide array of tasks, from cooking your food to lighting your dark hallways to charging your laptop battery. These devices also make a great place to get started if you still need some convincing (or just time to save up some money) before you embark on a larger solar installation. The following tools are a great way to expand your solar empire, or simply just start learning about the incredible amount of energy available for free right outside your window.

Figure 8.1. Using solar energy effectively means appropriate design. A clothesline, or solar clothes dryer, uses sunshine to accomplish the task of drying clothes—thus accomplishing a task that would take thousands of dollars worth of solar electric equipment to perform in an electric dryer. COURTESY OF LYNNE MONROE.

Solar Clothes Dryer

Far and away the cheapest and most effective way to use solar energy is a **solar clothes dryer**—of course, this is more conventionally called a clothesline. But don't let its simplicity or humdrum name fool you. Using a clothesline instead of an electric dryer can save a whopping 3.3 kWh of electricity each time. To put this in perspective, if you use your dryer once every day, to provide enough electricity to run your dryer in an average climate would require a 600 watt solar electric array. At an average of $10 per installed watt of solar electricity, that would cost about $6,000. A typical clothes drying setup of a retractable line, indoor rack, and an adequate number of clothespins would probably only set you back about $50 at the most. So a solar clothes dryer performs the equivalent task of $6,000 of fancy electronics! And your clothes will be fresher smelling for your trouble.

Cooking with the Sun

One of the most fun ways to utilize solar energy is with a **solar oven** or **solar cooker**. The premise of both is similar—using reflectors to concentrate solar energy on food, heating it up to temperatures of 350°F or more, plenty hot enough to do everything from cook stews to bake bread and even fry things like falafel!

Essentially there are two different methods of cooking with solar energy. The first type is generally referred to as a solar oven—as the name suggests it acts like a low-temperature oven (225°F–350°F), slowly cooking food over the course of several hours. A standard design mimics other solar thermal devices like a solar water heater or solar air heater—basically a glazed, insulated box pointed toward the sun, but in order to get temperatures hot enough to cook food, reflectors around the perimeter of the box concentrate the sunlight onto the glazing. Most models of solar ovens are portable, and are designed to be placed outside in a sunny spot, sometimes needing to be turned every few hours depending on the amount of cooking being done.

The second type of device is a solar cooker, and is parabolic in shape. A parabola is a shape that can concentrate incoming waves, such as the energy from the sun, onto a specific point, or focus. Solar cookers look more like a satellite dish than a typical solar collection device. Made out of reflective polished aluminum or some other mirrorlike material, the parabolic dish concentrates solar energy on a pot affixed to the focus of the parabola. They come in a range of sizes, from a two-foot diameter to a five-foot one. Unlike a solar oven, solar cookers are capable of bringing food up to temperature quickly, and can be used in the manner of a conventional stove top burner (assuming a clear sunny day), boiling water or oil for cooking things like pasta or frying food and vegetables like tempura and falafel. Obviously, larger diameter cookers are capable of heating foods up quicker and sustaining a higher temperature, and can be used earlier and later in the day. In sunny places like Colorado in

Figure 8.2. The Sun Oven has been cooking food with the sun all over the world for the last few decades. This tried and true cooker is the industry standard. COURTESY OF SUN OVENS INTERNATIONAL.

the summertime, we've used a five-foot diameter solar cooker to fry food at 7:30 in the evening.

The concept of concentrating solar energy onto a specific point to achieve high temperatures is one of the most exciting developments in solar energy, and creative people around the world are taking advantage of the concept to do an amazing variety of things. Solar Roast Coffee is a company that uses concentrating parabolic mirrors to roast coffee, achieving temperatures of 450°F–500°F and roasting green coffee to a smooth, rich brown in fifteen to twenty-five minutes. We'll explore the diverse applications of concentrated solar power (CSP) in more detail in the next chapter on "The Future of Solar."

Tried and True Solar Ovens and Cookers and Where to Get Them

We have three favorites when it comes to premanufactured solar ovens (see "Resources" for how to purchase them). The first two are conventional models that have stood the test of time. Both are manufactured by nonprofit organizations that help spread the word on solar cooking, especially in developing countries where solar ovens help combat deforestation and empower women by lessening dependence on wood and dung for cooking fuel. The first is the Sun Oven, manufactured by Sunovens International. It has been in production for over twenty years, and is a durable oven—easy to transport and fun to use.

The second tried and true model is the Sport Solar Cooker, which is less expensive at around $150 (and $25 more for the reflector), although a bit shallower. It is highly weather-resistant and sleek looking, and is an extremely effective solar oven.

The other model we've seen used to great effect is the Tulsi-Hybrid Solar Oven. It is mostly powered by the sun, but has a backup electric element to help maintain a constant temperature in case of cloud cover or a setting sun, making for more predictable cooking. It is also portable, but in order for the electric backup

Figure 8.3. The Sport Solar Cooker is another excellent cooker that will make a great addition to anyone's solar empire. COURTESY OF SOLAR OVEN SOCIETY.

element to work, a nearby outlet and extension cord are necessities. For the less sunny parts of the country, this can make a great compromise solar cooker. One of the drawbacks of these ovens is how shallow they are. Essentially, they look like a suitcase that has a mirror in the top section to reflect sunlight down into the glazed cooking area in the bottom section of the suitcase. Ordinary kitchen pots or loose food like winter squash do not fit well in here, unlike the much deeper Sun Oven or even Sport Solar Cooker. To account for this, the Tulsi-Hybrid comes with four shallow dishes. This shallowness can be awkward at times, but it also makes it possible to cook flat things like pizza or cookies well. Solar-cooked pizza! Yum! Another drawback is that since it has electric backup, the Tulsi-Hybrid should not be left out in the rain.

Obviously, you won't be using a solar oven to do all your cooking. But this doesn't mean you won't get loads of use out of it. And where kids are involved, solar cooking is a huge hit, especially if you're pulling out a tray of fresh-baked cookies! Because of their simplicity, portability, and effectiveness, they make great tools for teaching children and adults alike about the feasibility of solar energy.

Figure 8.4. The Tulsi-Hybrid Solar Oven uses backup electric heat to maintain temperature when the sun goes behind a cloud. A great tool for those of us who live in cloudier climes. COURTESY OF SUN BD CORP.

Figure 8.5. Here's Stephen cooking some falafel late in the afternoon on a parabolic solar cooker. Not as easy to find as more conventional solar ovens, but parabolic cookers are fun and useful because they can cook food early and late in the day.

Where solar ovens really come into their own, however, is cooking stews and soups, much like you would use a crock pot. Meat also does exceptionally well in a solar oven. The slow roasting means there is no burning to worry about, so you just stick your food in during the morning and pull it out cooked in the afternoon. Whole potatoes (sweet or regular) and winter squash also do great. Unlike things such as bread or cookies, stews and soups require no attention while they are cooking, so there's no need to worry about them while you're away at work.

We have less experience with different varieties of parabolic cookers, partly because there aren't as many available, and partly because we do most of our cooking in the evening when our solar window is shaded. When we've had the pleasure of using a parabolic cooker, it's been awesome, and we still have a hankering for one even though they're of limited utility for us. If you have late afternoon/early evening access to the sun, or if you cook regularly during the day, then a parabolic cooker can be a great addition to your solar empire. One company that's been around a while and makes two types of parabolic cookers is Mueller Solartechnik in Germany. Cookers can be ordered directly from their Web site. See "Resources" at the end of the chapter.

Lighting the Home and Cooling the Attic

These next two easy solar items are inexpensive and very effective but typically require installation in your home by a professional. These are tubular daylighting devices and solar-powered attic vents. We group these together because they both are installed on the roof. They both work without having to fuss about them, but you'll want to position them out of the way of any large arrays of other solar panels that might need plenty of roof space in the future. Often both of these devices function well even when placed on non-south portions of a home's roof, but inquire with the installer or manufacturer if you have any doubts.

Daylighting

Many homes have rooms or hallways that don't receive adequate light during the day, meaning it's almost always necessary to turn on an electric light when people enter them. Having to use fossil fuel-powered light when the sun is shining outside doesn't make any sense at all. The installation of a daylighting device can potentially solve this problem, at least on the top floor of a house. The principle is quite simple; they're basically a periscope. A small dome sticks out above the roof line taking in sunlight, which is then reflected down a tube into the home. Where previously a skylight would require a reworking of roof rafters and ceiling joists, a tubular daylighting device can fit between existing rafters and ceiling joists. The hole made in the ceiling is covered with a base like a more conventional ceiling light—rather than having to frame in a large opening in the ceiling as with a skylight, and finish it off with drywall and paint. This makes installation much, much easier. Heat loss is increased slightly (way less than with a skylight, however), but the additional lighting more than compensates for this loss of heat energy-wise. A typical installation will cost five hundred dollars or so.

Daylighting retrofits are not limited to tubular devices on the top floors of buildings. Several companies are making daylighting systems that use fiber optics to transport the sunlight as much as

Figure 8.6. The Solatube Daylighting Device brings in lots of light without the hassles and hard work of a skylight (and with much less worry about leaks). COURTESY OF SOLATUBE.

four floors down. These systems are expensive, starting at ten thousand dollars without installation (so they're not exactly cheap and easy)—but they might make sense for larger retrofits. Since they can provide light for up to one thousand square feet, they can bring daylighting into large loft spaces. Hopefully once the technology becomes established, the price will come down—although installation difficulties will almost certainly keep the price relatively high.

Cooling the Attic

Solar attic fans are a great addition to your roof in warmer parts of the country. Roofs exposed to long periods of sun in the summertime can get to over 180°F, and it's inevitable that some of this heat will work its way into your attic and seep through your ceiling insulation into your home. The more you can keep this heat from building up in your attic, the less it will make its way into your

Figure 8.7. Overheated attics can push a tremendous amount of heat into the living space. A solar attic fan can do a great job cooling down your attic, but make sure not to place it where it might interfere with later solar installations.

home, and the less time your energy-intensive AC system will have to run. Solar attic fans do just that. And because their power source is the sun, they only work when they need to. Having their own self-contained energy source, installation is straightforward and doesn't require an electrician. The typical installation cost as of this writing is around one thousand dollars or so.

One thing that's interesting to note is that many of the larger solar installations—whether electric, hot water, or heating—have the benefit of shading the roof and helping bring attic temperatures down, perhaps making a solar attic fan unnecessary. This is something to consider if you think you might invest in any of these solar systems in the near future. If you're going to have a daylighting device or attic fan installed, try to bear in mind any solar systems you may want to add to your home further down the road, even if it's not in the immediate future. Where possible, avoid having these devices placed near the center of large open spaces on your south-facing roof, where they could potentially interfere with other solar installations.

Solar in the Yard

Probably most people's introduction to solar electricity comes from solar-powered patio or walkway lights. Where only small amounts of light are needed, especially to guide the way to the front door on a dark night, solar patio lights are an extremely easy gateway to solar electricity. They're basically little off-grid PV systems, all self-contained and ready to install for as little as fifteen dollars! One of the reasons that they're so inexpensive is that they use the cuts from the squaring of round PV crystals (see the sidebar "Lifting the Wizard's Curtain" in Chapter 5 on how PV modules are made). The biggest mistake we see with these is putting them along paths in wooded areas—these are solar devices and won't do diddly-squat if they don't get several hours of direct sun every day.

Figure 8.8. Solar yard lights are actually little self-contained off-grid PV systems. Not bad for under twenty dollars!

Figure 8.9. Rechargeable batteries not only can be recharged using solar energy, but they also don't produce the toxic waste that single-use batteries do.

Distillation and Food Dehydration

The two topics of water distillation and food dehydration are important to mention even though readily available premanufactured models of solar water distillers and food dryers are either expensive or don't exist. Water and food are important, right? And solar energy can be used very effectively to ensure access to these necessities.

Solar energy is capable of distilling dirty water and making it potable. The principle is simple: water is heated, evaporates, and condenses on a surface where it collects, rolling down into a separate sterile container when enough has accumulated. Such applications make sense where access to potable water can be a challenge, such as on a long boat voyage or in areas subject to water supply disruptions due to drought, hurricanes, or other catastrophic events. Premanufactured solar distillers tend to be expensive, catering to the boating crowd, but the concept is simple—making an effective distiller out of clear bottles is possible.

For the avid gardener, a simple method of preserving excess bounty is with a solar food dehydrator. Only one premanufactured model exists, and its design leaves something to be desired, especially compared with the standard home-built model as described by Eben Fodor in his book *The Solar Food Dryer*. We would especially like to see a well-built premanufactured model that has an electric backup, similar to the Tulsi-Hybrid Solar Oven, to ensure that any solar dehydrating is brought to completion even if the sun goes away for a few cold, rainy days.

Moving Water: Irrigation and Fountains

When the sun is out, plants need water. Because solar electricity generates power when the sun is shining, it is often used for moving water from reservoirs to thirsty plants by farmers and gardeners alike. This is generally done by powering a DC pump directly with a photovoltaic module (i.e., solar panel). Likewise, water can be moved for pleasure in the form of landscaping fountains. If you're willing to only have the pleasant sound of gurgling water when the sun is shining (and when most people are likely to be hanging out in the yard), then a fountain can be powered directly by a DC pump. For continuous gurgling, batteries are needed to store power for less sunny times.

Solar Battery Chargers

One other solar-powered item we can unequivocally endorse is a solar battery charger. Small, used-up (nonrechargeable) batteries are toxic waste, and our society has never really come up with an effective method of disposing of or recycling them. We frequently see them littered on the ground and tossed casually in garbage cans. By one measurement, improper disposal of household batteries accounts for 88 percent of the mercury found in a municipal waste stream. Using rechargeable batteries instead of disposable batteries can save several hundred if not several thousand batteries from the garbage can (higher end rechargeables are good for up to three thousand charges). You can buy battery chargers that plug into a wall socket, or have a solar panel embedded in the top as a charging source. Another possibility is a solar-powered battery charger built specifically for everyday electronics like your smart phone or laptops. They work just fine hanging out in a south-facing window, and some are even embedded in backpacks.

Other "Solar" Appliances

Remember, using solar energy effectively is about collecting meaningful quantities of dispersed energy over substantial lengths of time. It has become all the rage lately to slap a PV module on just about anything and call it "green," "sustainable," or "solar." Flipping through some catalogs and perusing the Internet reveals some items of high quackery—such as a solar-powered water bottle, a rotating

Figure 8.10. Lots of folks have tried to jump on the bandwagon by sticking a "solar" in front of their product's name. This particular product doesn't have the slightest thing to do with harvesting solar energy, and it looks mighty uncomfortable.

solar plant holder, and a solar cigarette lighter, to name just a few. No, the water bottle does not purify water (it just has a light inside of it adding weight and taking up space), last we checked the earth itself was a rotating solar plant holder, and we'd guess it's more likely for someone using a solar cigarette lighter to burn out their eyeballs than get their smoke lit.

Perhaps taking the cake in useless solar devices are the solar vinyl shorts pictured in Figure 8.10. Essentially a rejiggered garbage bag, wearing the solar vinyl shorts results in a "dehydration process" that "helps rid your body of excess water weight and shed unwanted inches." Essentially, part of your body is in a black garbage bag and you're rapidly losing water due to sweating. Guess what? In order to continue to live, you'll need to drink the equivalent amount of water that the solar vinyl shorts so helpfully rid you of.

Such solar gimmickry is generally easy to detect. Solar energy is amazing stuff, and capable of powering much of our lives, but it's important to keep the challenges of using solar energy effectively in mind. The devices and gadgets outlined in this chapter are legitimate (except for the shorts!), but as with any other time you're asked to open up your wallet, take your time in deciding and check around with others to see if the product is effective. *Caveat emptor!*

Resources

The Sun Oven costs around $300: http://www.sunoven.com
Phone: (800) 408-7919

Mueller Solartechnik Parabolic Solar Ovens:
http://www.mueller-solartechnik.com

The Sport Solar Cooker can be ordered direct from the Solar Oven Society: http://www.solarovens.org
Phone: (612) 623-4700

Tulsi-Hybrid Solar Ovens: http://www.sunbdcorp.com

Solatube: http://www.solatube.com

CHAPTER 9

The Future of Solar

Our homes and lives require prodigious amounts of energy, with only a miniscule amount being supplied by solar energy. The earth receives vast quantities of solar power, much more than the world currently uses, so theoretically every human activity has the potential to be run on sunshine. From powering our cars to fueling industry, the scope of solar applications is growing by the day, restricted only by the design of our infrastructure and the creativity and dedication of our leaders.

What are some of the cutting-edge breakthroughs that are occurring right now, and how might the solar landscape look a decade or two hence? As physicist Niels Bohr realized, prediction is difficult, especially about the future. Yet there's much truth in novelist William Gibson's quip that the future is here, it's just not widely distributed yet. Tapping the power of the sun has intrigued humanity for millennia, and does so now more than ever—from the backyard tinkerer perfecting his homemade solar air heater, to the scientist in the lab working on solar collection devices on the atomic scale.

The bulk of today's solar research budget goes toward improving the efficiency and cost-effectiveness of photovoltaics. Only about 6 to 20 percent of sunshine falling on the typical PV module is converted into electricity, so there's some room for improvement. This is in contrast to solar thermal devices where the technology is relatively straightforward and efficiencies are already high (50 to 70 percent) and thus harder to improve.

But this desire for increased PV efficiency has led to no small amount of solar monomania that neglects the importance of using solar energy for what it does best: heating things up, specifically solar thermal applications like solar hot water and solar heating. There is also a fair amount of hyperbole about solar electricity that gets bantered around in the news about imminent breakthroughs in PV, and this can unfortunately create a wait-and-see attitude in regard to all things solar. As we've said before, all the solar technologies covered in this book work amazingly well right now. For the homeowner, the best thing to do is concentrate on the option with the highest efficiency that is most cost-effective for your particular situation based on your needs, budget, and solar resource. With a little bit of dedication on our part, our children and grandchildren (and every generation thereafter) will enjoy a solar-powered existence in harmony with the planet.

Large-Scale Solar Thermal Technologies

The much greater efficiency of using solar energy for heating rather than electricity is demonstrated by concentrating solar power (CSP) electric-generating plants. These are utility-scale power plants located in sunny areas that use mirrors to concentrate solar energy onto a fluid. The heated fluid then spins a turbine and generates electricity, in much the same way as traditional power-generating equipment at coal- or natural-gas-fired power plants. In some CSP plants, thermal storage of the solar energy allows

Figure 9.1. Photovoltaics aren't the only way to make electricity from sunlight. On a large scale, concentrating solar thermal energy and using it to heat a liquid that then turns a turbine may prove equally efficient. Several of these concentrating solar power plants are being built today. PHOTO BY K. J. KOLB.

Figure 9.2. Large-scale solar electric plants are being built throughout the country, and promise to supply the grid with substantial quantities of renewable electricity. COURTESY SOUTHERN ENERGY MANAGEMENT.

electricity generation well into the evening hours, when demand is often highest—overcoming a hurdle presented by conventional PV generation.

The uses of CSP are limited only to the size of the collecting device, typically a parabolic dish. The parabola can be expanded almost indefinitely to reach higher and higher temperatures. A solar furnace, built in Odeillo, France, in 1970 reaches temperatures exceeding 6,000°F and demonstrates the potential use of solar energy in industrial manufacturing applications.

Solar thermal technologies have vastly underused potential in heating all types of commercial buildings as well as residences. Every large building that has a significant heating load, from stores to offices to factories, has the ability to use solar energy for preheating the air used in keeping the building warm. A great example of preheating air comes from a company called SolarWall. When the sun is shining, air is preheated through black metal channels before entering the conventional HVAC equipment.

Large-scale solar electric farms are also sprouting up all over the country. As incentives remain strong and incremental improvements in the components of these systems (not just the PV modules themselves but also the inverters, racking, and other equipment) continue, the financial prospects for large-scale PV will continue to entice all types of investors. This should mean a solar electricity generating plant near where you live in the near future, great news for potentially tapping into carbon-free renewable electricity through green-power programs for those who lack the solar or financial resources to make it happen on their own buildings.

Residential Applications

Regarding residential applications of solar energy, it is extremely unlikely that there will be new technologies available for the average home or office before the end of 2016 when the federal tax credit is currently set to expire. The solar thermal technologies of heating water and buildings are already very efficient at 50 to 70 percent, so there's no reason to wait. Even if some unlikely breakthrough in PV technology suddenly doubled the efficiency from an average of 15 to 30 percent, not only would market availability remain a decade away for new products incorporating the higher efficiency, but there are also many other costs associated with a solar electric installation. Even if such a discovery halved the module's cost (by no means a foregone conclusion) it would only reduce the price of the installed system by about a quarter, since typically the PV modules themselves are about half the cost of any installation (the rest going to labor, copper wire, inverters, racking, and so on). This 25 percent reduction is less than the federal tax credit of 30 percent.

That's why we believe the future of solar is now! Solar technology has been improving for hundreds if not thousands of years. It's time to take the plunge. If there are areas where solar needs to be improved, it's more in the legislative sphere than in the lab. Incorporating fossil fuels' finite and polluting nature into their cost is the most obvious legislative turn-around we need (hopefully something will be done in this regard by the time you read this). Financial incentives that vary with shifts in political power can create boom/bust cycles that are hard on a burgeoning solar workforce, so enacting permanent laws that price renewable energy *and* fossil fuel energy according to their true costs (i.e., incorporating external costs like fossil fuels' pollution, dependency, and finiteness) will make for a steady and consistent conversion to solar and

Figure 9.3. At long last, electric cars are becoming widely available, making the possibility of solar-powered transportation no longer just a pipe dream.

other renewable power. Working out solar access laws, net metering policies, permitting, and utility interconnection standards in a comprehensive and standard way is also a dire necessity for removing unnecessary complications to customers and installers alike.

In the future we see all homes utilizing solar power for much if not all of their energy needs, including transportation needs like the car or electric bike out in the garage or the commuter train stopping just down the block. Probably more important than intriguing technologies like solar nanotechnology or spray-on-PV in the long term is the possibility of energy- and cost-efficient storage of electricity. Smoothing out the availability of solar energy through efficient storage may become possible through hybrid or electric car batteries tied into the grid or in distributed fuel cells (most likely stationary ones in the home or neighborhood). Such technology would eliminate worries about solar energy's variability, and free up the possibility of powering every last bit of our civilization using solar energy.

In the meantime, we've got a lot of work to do making use of the many tried and true solar technologies described in this book. The future of solar is now, but only if each of us as individuals make it a priority in our lives. Good luck on your journey to a solar-powered existence, and make sure to keep your head held high as you walk toward the sun!

More Reading

Books

The Low-Carbon Diet by David Gershon
The Carbon-Free Home by Stephen and Rebekah Hren
Plan B: 4.0 by Lester Brown

Magazines

Solar Today
Homepower

Internet

Solar News: http://www.solarbuzz.com
Renewable Energy News: http://www.renewableenergyworld.com
Solar Energy Industries Association: http://www.seia.org
American Solar Energy Society: http://www.ases.org
Energy and Sustainability News: http://www.energybulletin.net

Index

About the Authors

Stephen and Rebekah Hren live in Durham, North Carolina, where they are both actively involved with renewable energy, natural building, and edible urban gardening. They are the authors of *The Carbon-Free Home,* and both frequently write articles and speak on renewable energy and sustainable building.

Rebekah is a NABCEP-certified solar photovoltaic installer, licensed electrical contractor, and ISPQ-certified solar instructor for Solar Energy International. She designs and installs PV systems of all sizes, from small off-grid systems to utility-scale arrays, and teaches photovoltaic design classes across the country.

Stephen is a builder and teacher with experience in sustainable design and passive and active solar-heating technologies. He works as a professional restoration carpenter.

Find out more about the Hrens at www.thecarbon freehome.com.

Chelsea Green Publishing is committed to preserving ancient forests and natural resources. We elected to print this title on 30-percent postconsumer recycled paper, processed chlorine-free. As a result, for this printing, we have saved:

8 Trees (40' tall and 6-8" diameter)
3 Million BTUs of Total Energy
755 Pounds of Greenhouse Gases
3,636 Gallons of Wastewater
221 Pounds of Solid Waste

Chelsea Green Publishing made this paper choice because we and our printer, Thomson-Shore, Inc., are members of the Green Press Initiative, a nonprofit program dedicated to supporting authors, publishers, and suppliers in their efforts to reduce their use of fiber obtained from endangered forests. For more information, visit: www.greenpressinitiative.org.

Environmental impact estimates were made using the Environmental Defense Paper Calculator. For more information visit: www.papercalculator.org.